FIREFIGHTER SUCCESS

Stay Fit, Stay Safe!

20 C's TO FIREFIGHTER EXCELLENCE

JIM MOSS

Copyright © 2020 - Firefighter Success, LLC

All rights reserved. No part of this book may be reproduced, stored in a retrieval system, or transcribed in any form or by any means, electronic or mechanical, including photocopying and recording, without prior written permission of the publisher. Brief quotations and photos of the book and its content may be shared via digital/social media and in book reviews, provided that *Firefighter Success* is attributed as the source.

Requests for permission to reproduce any part of this work should be directed to the publisher at: FirefighterSuccessBook.com

This publication is designed to provide accurate and authoritative information in regards to the subject matter covered. It is sold with the understanding that the publisher is not engaged in rendering legal advice or other professional services.

Publisher: Firefighter Success, LLC
FirefighterSuccessBook.com
Instagram | Facebook - @FirefighterSuccess
Twitter - @FireSuccessBook

Author - Jim Moss
Introduction - Mark vonAppen
Foreword - Jason Hoevelmann
Edited by - Patrick Murphy
Technical Editor - Dan Kerrigan
Photographer - Chris Smead
Book Design & Layout - Dave Chojnacki

Library of Congress Cataloging-in-publication data available upon request.

Firefighter Success / Jim Moss

Printed Paperback
ISBN - 9798621486129
Printed in the United States of America

PRODUCED AND PUBLISHED BY

WHEN LIVES DEPEND ON US, SUCCESS IS OUR ONLY OPTION.

A portion of each book purchase is donated to first responder charities.

TABLE OF CONTENTS

ACKNOWLEDGEMENTS...6
ENDORSEMENTS...7
INTRODUCTION...13
FOREWORD..17
FROM THE AUTHOR..21

CHAPTERS

1 - COACHABLE...25
2 - CURIOUS..35
3 - CHAMPION MINDSET...43
4 - CONFIDENT...51
5 - COMMITTED...57
6 - COURAGEOUS..69
7 - CONVICTION..83
8 - COMPETENT...91
9 - COMPREHENSIVE..109
10 - CONSISTENCY...117
11 - CHARACTER..127
12 - CREDIBILITY..137
13 - CHARISMA...145
14 - COMMUNICATION..149
15 - CANDID..163
16 - COMPASSION..169
17 - COMPOSURE...177
18 - CHANGE...185
19 - COMMUNITY...201
20 - COACH...211

ADDITIONAL RESOURCES..219
ABOUT THE AUTHOR...225
ABOUT THE PHOTOGRAPHER..227

ACKNOWLEDGEMENTS

TO GOD:

For your overabundant provision in every area of my life. I am forever grateful for your forgiveness and salvation. It is only by your grace that I have made it this far. Use me to do your will.

TO ALI:

For your never-ending support, patience, and encouragement. I am grateful for your love, which inspires me to be a better husband and leader. You are the best teammate I could have ever asked for.

TO IZZY, MAGGIE, AND JIMMY:

For your love, joy, and curiosity. You motivate me to be the best dad I can be. May you "grow in wisdom and stature, and in favor with God and man."

TO MY MENTORS:

For passing your wisdom and experience on to me. I am grateful for the most precious gift you have given me: your time. I hope to make you proud and to leave a legacy worth following.

ENDORSEMENTS

Becoming a firefighter is a choice, and it's not for everyone. It's about leading from every level; it's about defeating mediocrity and entitlement—every day and in every way. It's about them, not you.

In *Firefighter Success*, Jim has given you 20 core principles that, if applied, will not only help you excel at your job but also in life, because they are more than just words. This book contains life lessons and experiences, both good and bad, that come together to form a message of excellence—which is shared from a place of humility and gratitude. I guarantee you that if you adhere to these principles, you will not only realize personal success but you will also serve as an inspiration and example to those around you as well.

The fire service needs people to do the work worth doing ... the hard work ... the work that separates the great from the good, the innovators from the status quo. There is no secret formula for success. It starts and ends with you. Decide. Commit. Work hard. If you are up for the challenge, *Firefighter Success* will go a long way toward getting you there.

Dan Kerrigan
Chief - Township of Upper Providence Department of Fire and Emergency Services (PA)
Bestselling Co-author - *Firefighter Functional Fitness*
FirefighterFunctionalFitness.com

Firefighter Success: 20 C's to Firefighter Excellence takes us on a comprehensive journey and gives us the tools to attack the challenges we'll encounter. It lays the foundation for long and truly rewarding lives; both in and outside the fire service. In his 20 C's, Jim pours out his heart and passion for the job, helping us realize that our greatest personal success comes solely from helping others to be successful.

Jim Moss is one of the most sincere, caring, and giving people I know in this vast fire service. He embodies what I have long promoted: *"It's not about us: It's about those we serve, those we serve with, and most importantly, those who allow us to serve: our predecessors, our leaders, and our families, those who love us most."*

Tiger Schmittendorf
Deputy Fire Coordinator (Ret.) - Erie County Department of Homeland Security & Emergency Services (NY)
Vice President of Strategic Services - FirstArriving.com
TrainYourReplacement.com

Dedicated, committed, reliable, tough, educated, focused, and coachable—those are the qualities I look for in firefighters. They also happen to be the traits my friend Jim Moss possesses, which is why I highly recommend *Firefighter Success*. If you want to become a great firefighter, read it and live it.

Frank Viscuso
Deputy Chief (ret.) - **Kearny Fire Dept. (NJ)**
Bestselling Author - *Step Up and Lead*
CommonValor.com

In *Firefighter Success: 20 C's to Firefighter Excellence*, Jim gives an honest, straightforward, and humble approach to grassroots leadership and personal development. He not only helps you build a rock-solid foundation to improve as a firefighter and leader but, more importantly, also to be a better person. The formula is easy: read, learn, apply … and repeat.

Paul Combs
Lieutenant (ret.) - **City of Bryan Fire Dept. (OH)**
Cartoonist - *Drawn by Fire*
ArtStudioSeven.com

If you want to be a firefighter who knows the importance of work performance and professionalism, *Firefighter Success* will provide you with critical principles to guide you from building your foundation as a firefighter to growing into one that others look up to and want to emulate.

Dr. David Griffin
Battalion Chief - **Charleston Fire Dept. (SC)**
Bestselling Author - *In Honor of the Charleston 9*
DrDavidGriffin.com

Firefighter Success provides a positive road map for firefighters within all levels of the fire service. In this book, Jim has compiled and shared profound core values and insights that will ensure a fulfilling career.

Eric Wheaton
Lieutenant - Winter Park Fire Dept. (FL)
Owner - Vent Enter Search LLC
VentEnterSearch.com

Humble, confident, realistic, and more. Simply put, *Firefighter Success* is outstanding! Jim hits it out of the park with this book. This is a great read that will help every firefighter from the bottom of the organizational chart to the very top. It will inspire you to be a better version of yourself both professionally and personally.

Jim uses his knowledge, experience, success, and faults to share his hunger for the development of firefighters everywhere. It's easy to see his passion pour out of every page.

Jarrod Sergi
Captain - Norfolk Fire Rescue (VA)
Author - *No Nonsense Leadership: A Realistic Approach for the Company Officer*
TrialByFireOnline.com

From the probie to the 30+ year veteran, this book is a must read for every thriving firefighter. These pages are packed with value and essential information that clearly come from years of experience and service. *Firefighter Success* educates the reader on how to make the most of this amazing profession. Jim has provided the road map for success in this book that should not only be adopted but also be utilized as a reference throughout the rest of every firefighter's career.

Steve Hughes
Firefighter - Edmonton Fire Rescue Services (Canada)
Director - Train Your Probie Inc.
TrainYourProbie.com

Firefighter Success has a double-sided approach: It teaches all firefighters a solid foundation in the basics while also pushing them to achieve excellence in their careers. This will be a book you can read and re-read throughout your career to help you be your best.

Blake Stinnett
Firefighter - Sandy Springs Fire (GA)
Founder/Director of Next Rung - NextRung.org
Co-founder - Open Bale

Firefighter Success gives all firefighters, regardless of rank or years of experience, the essential concepts to grow and succeed in every aspect of life. By following its principles, you will benefit in all areas of your life.

Devon Wells
Fire Chief (ret.) - Hood River Fire & EMS (OR)
Founder - The Rural Fire Officer

Firefighter Success is an incredible resource for anybody looking for that spark of motivation to reignite their passion, or a new firefighter looking for solid guidance that will lead to career success. An important concept Jim returns to throughout this book is the power of overcoming failure. It's not a book about striving for perfection, but rather achieving excellence through proper mindset and attitude. For anybody looking for a few tips for overcoming challenges and obtaining success, this book won't let you down.

I have been fortunate to know Jim for the last several years, and from the first day I met him, he was a breath of fresh air. He always has a smile and kind word to share. Of all that he has to offer, his genuine nature and humility are his greatest gifts. Jim's message is a one that he lives and exhibits day in and day out.

Dena Ali
Captain - Raleigh Fire Dept. (NC)
Founder/Director - North Carolina Peer Support

The fire service has a rich and colorful history of dedicated individuals whose lives have greatly impacted their generation and those who followed. Successful firefighters who have left legacies knew one thing above all else: Their service and what they gave to others were far more important than the servant.

In his book, *Firefighter Success*, Capt. Jim Moss breaks down the values of successful firefighters who came before us. As you read this book, I challenge you to do as I have: Look upon each chapter as an opportunity to explore each principle and how we can live them out in our daily lives. Whether we realize it or not, our examples are shaping the next generation and the future of the fire service. The question we must ask ourselves: Are we living these qualities out to the fullest?

Andrew J. Starnes
Battalion Chief - Charlotte Fire Dept. (NC)
Founder/Owner - Insight Training LLC
InsightTrainingLLC.com

Jim Moss is someone who has lived out the principles in *Firefighter Success*. It has proven to lead him to success and it will do the same for you when applied.

John Spera
Firefighter - South Metro Fire Rescue (CO)
Founder - FitToFightFire.com
Author - *Mindset*

Firefighter Success is a great compilation of timeless principles and traits for firefighters. When implemented, they will produce success in the fire service and in life. It is easier said than done, but those who accept the challenge and exemplify these character traits will become a successful firefighter and leader.

David J. Soler
Founder - Firefighter Toolbox
Bestselling Author - *Firefighter Preplan: The Ultimate Guide For Thriving as a Firefighter*
FirefighterToolbox.com & FirefighterPreplan.com

Being a fire academy instructor for the St. Louis Fire Department for more than 20 years, I have taught the technical and practical aspects of the job to aspiring firefighters. During my time with these recruits, I have also interjected personal leadership principles to help them maximize their success. Over the past 10 years, I have spread my *GLUE Personal Leadership* message to firefighters on both the national and international levels.

I couldn't be more happy for my fire service brother Jim Moss. He epitomizes the 20 C's found in *Firefighter Success*. In this book, Jim shares how every firefighter can achieve personal greatness, and skillfully shows how simple, everyday personal investments chart the path to success.

If you are striving to be the best firefighter you can become, then I'm sure you have your list of go-to books for guidance and motivation. Make room on your bookshelf because *Firefighter Success* is that next book in your arsenal. Enjoy the read and the growth that follows!

Larry Conley
Captain - St. Louis Fire Dept. (MO)
Leadership Development Concepts, LLC
GlueNationLDC.com

INTRODUCTION

Capt. Mark vonAppen has been a member of the Palo Alto, California Fire Department since 1998. He is currently assigned to the suppression division. He is a committee member for California State Fire Training and has contributed to the development of Firefighter Survival and Rapid Intervention curriculums. Mark has been published in *Fire Engineering*, *Fire Service Warrior*, and is the creator of the fire service leadership movement Fully Involved. He is an instructor for the Santa Clara County Joint Fire Academy, a recruit instructor for Palo Alto Fire, an academy instructor at Evergreen Community College, and a member of the Nobody Gets Left Behind training group.

The legendary basketball coach of the UCLA Bruins, John Wooden, was a stickler for routine. Wooden coached The Bruins to 10 national championships from 1948-1975, and they didn't happen by accident. The streak of championships included seven consecutive titles, a feat that might not ever be equaled.

Wooden believed so deeply in the power of preparation that he left no detail to chance. To some, his style may have seemed maniacal and all-consuming, but his methods achieved unimaginable results. At the beginning of each season, Wooden would instruct all of his players, from the high school recruit to the most decorated senior, on how to properly put on their socks and shoes.

> *"I think it's the little things that really count. The first thing I would show our players at our first meeting was how to take a little extra time putting on their shoes and socks properly. 'The most important part of your equipment is your shoes and socks. You play on a hard floor. So you must have shoes that fit right. You must not permit your socks to have wrinkles around the little toe; where you generally get blisters, or around the heels.'*
>
> *"It took just a few minutes, but I did show my players how I wanted them to do it. 'Hold up the sock, work it around the little toe area and the heel area so that there are no wrinkles. Smooth it out good. Then hold the sock up while you put the shoe on. The shoe must be spread apart; not just pulled on by the top laces. You tighten it up snugly by each eyelet. Then you tie it. Then you double-tie it so it won't come undone; because I don't want shoes coming untied during practice or during a game.'*

INTRODUCTION

> *"I'm sure that once I started teaching that, years ago, that we cut down on blisters. It definitely helped. But that's just a little detail that coaches must take advantage of, because it's the little details that make the big things come about."*[1]
>
> -Coach John Wooden

Success in the fire service, as in sports, is no accident. And success in both arenas is measured by our wins and losses. As a firefighter, achieving it is a continuous process that requires an unwavering dedication to the details. And it doesn't come without deliberate planning and preparation.

Each of us must have the Coach Wooden mindset about "the little things," which includes putting on our socks and pulling up our boots properly. Whether we are the rookie or the senior firefighter, we will own our roles and responsibilities, down to the smallest detail. For the former, this may mean learning how to turn out and mask up quickly, and, for the latter, it might mean mentoring the next generation. Wherever we are in our journeys, it is critical that we first invest in ourselves if we are to be the best firefighter and team member that we can be. In essence, we must first lead ourselves.

Firefighting is a team sport, and our team must be able to fully depend on us. Success is determined by whether or not our teammates can count on us to deliver. Trust, reliability, and our individual contributions are what will ensure the team's success. As a team, we all must see things through the same eyes; believing in the oneness that is created through a grinding dedication to the mission and to one another. It is through a mutually agreed upon belief system that guarantees everyone's success.

I first came to know Jim for his role with *Firefighter Functional Fitness*, where his passion for helping firefighters was obvious. As I have gotten to know him more, I will tell you that he is someone who exemplifies leadership, ownership, and character. All of the principles that he shares in this book aren't merely theory or suggestions—he lives them out on a daily basis. Jim is a man of action who leads by example—exactly what the fire service needs.

With *Firefighter Success*, Jim provides a step-by-step guide to help every firefighter prepare for a successful and rewarding career in the fire service. Its 20 principles are the foundation to every firefighter's journey to excellence. The simple, yet timeless wisdom contained within provides the details to winning as a firefighter.

Don't just read this book and put it on the shelf. Use it as motivation. Use it to develop self-discipline. Use it to build the habits of excellence which all successful firefighters possess. Most importantly, use it as a guide to becoming the best firefighter that you can possibly be. Absorb it, own it, and live it.

---

1 "John Wooden: First, How to Put on your Socks." *Newsweek*, Oct. 24, 1999, www.newsweek.com/john-wooden-first-how-put-your-socks-167942.

Finally, and most importantly:

Do your job.
Treat people right.
Give all-out effort.
Have an all-in attitude.

Mark vonAppen
Captain - Palo Alto Fire Department (CA)
Founder and Owner - Fully Involved
Mark-vonappen.blogspot.com

FOREWORD

Jason Hoevelmann is the fire chief of the Florissant Valley Fire Protection District and a volunteer deputy chief/fire marshal with the Sullivan Fire Protection District in Sullivan, MO. He is the author of *No Exceptions Leadership* and *The New Company Officer*. He has been a fire service instructor for 25 years and has a bachelor's degree in Fire Service Administration from Eastern Oregon University. Jason is a member of the International Society of Fire Service Instructors and serves on the board of directors of the International Association of Fire Chiefs' Fire-Life Safety Section and the Company Officer Section. He is also a contributor to *Fire Engineering* and *Fire Rescue* and shares his leadership message at fire departments and conferences throughout North America.

When Jim asked me to contribute to this book, I was incredibly humbled and honored to do so. On a personal level, it is inspirational to see how Jim has risen through the ranks of his organization and also become a leader in the fire service. And it is gratifying to see that he authored a bestselling book that has impacted the health and well-being of so many firefighters across the globe.

I was not surprised, however, to read the great information Jim has shared in this book. As I read through it, all I could keep thinking about was how valuable his message is for future and current fire officers as they navigate the choppy waters of leadership. *Firefighter Success*' content and format make it easy to read and then reference later on—whether you need specific information or just a little encouragement for a difficult situation.

The 20 core principles Jim has provided are the perfect format for covering a variety of situations that you are sure to encounter as a firefighter and fire officer. Being able to frequently refer back to these principles to evaluate your current situation is invaluable. With the *Action Steps* found at the end of each chapter, this book is a complete officer development guide for beginners and veterans alike. Additionally, these *Action Steps* can be used by company officers to develop their future officers. What an enormously valuable tool!

My hope is you find the value in this book that I did and that you will refer to it often. Share its information with everyone: *aspiring firefighters, rookies, veterans, future and current officers, etc*. Information is best when it is shared with those who desire to know more and those who want to improve. Leadership and personal development are not finite—they require the pursuit of excellence, which is a continuous journey. Just like

FOREWORD

any other skill, both are perishable.

Firefighter Success is the go-to guide to help you stay on the right path to a career of firefighter excellence.

Jason Hoevelmann
Fire Chief - Florissant Valley Fire Protection District (MO)
Author of *No Exceptions Leadership and The New Company Officer*
Founder and Owner - Engine House Training, LLC
EngineHouseTraining.com

WHEN LIVES DEPEND ON US, SUCCESS IS OUR ONLY OPTION.

FirefighterSuccessBook.com

FROM THE AUTHOR

Jim Moss
Author of *Firefighter Success*

FirefighterSuccessBook.com
Instagram | Facebook: @FirefighterSuccess
Twitter: @FireSuccessBook

> "Doing the best we can with what we are given—this is the true definition of success."

First and foremost, I am extremely grateful this book is in your hands. Whether you purchased it for yourself or it was gifted to you, it will be a worthwhile investment of your time. Perhaps you have 30 years on this job or maybe you only have 30 days as a firefighter … either way, I am sure that you will find its principles to be helpful and inspiring for the rest of your career.

I have poured my passion into every sentence of this book, with the singular mission of helping every firefighter achieve excellence in their careers and in their personal lives. It is my honest desire to see every firefighter succeed—*and that is why I wrote this book.*

I did not "invent" *Firefighter Success*' principles. I have learned them from successful firefighters, and from successful individuals outside of the fire service. I have also learned from my experiences and especially from my mistakes. I have been inspired by many areas of my life: my faith, my mentors, my family and friends, and more.

I will be the first to admit that I am not a perfect example of everything contained in this book. Just like you, I am a work in progress on my journey to continuous improvement and success.

WHAT IS SUCCESS?

If we surveyed 100 firefighters to ask them their definition of success, we would most likely get 100 different responses. However, there is one belief we must all have in common: When lives depend on us, success is our only option.

> **For the purpose of this book, firefighter success is defined as:**
> - Achieving excellence in every aspect of life: as a firefighter, in our personal lives, with our families, etc.,
> - Striving for continuous self-improvement and growth,
> - Maximizing and achieving our potential,
> - Doing the best we can with what we are given,
> - Leaving a positive, lasting impact on others, and
> - Refusing to give in to failure.

As fire recruits and new firefighters, we are typically only taught the *skills* of the job. But becoming a successful firefighter is so much more than just the job performance requirements. Achieving excellence, growth, and maximizing our potential goes far beyond basic competencies and technical knowledge. As firefighters, our long-term success is dependent on the 20 C's found in this book: *character, confidence, courage, compassion, etc.* (just to name a few). When we adopt *Firefighter Success'* principles and live them out through action, we build rock-solid foundations to incredibly rewarding careers and lives.

> "We are much more than the sum of our skills, strategies, and tactics."

Everyone's journey to success is unique, and we must know the process isn't one-size-fits-all. All of us have different strengths and weaknesses; therefore, we each have multiple areas for improvement and growth. No matter what we do or how slow we go, we will always move forward. We will focus on making progress and becoming the best that we can possibly be.

SUCCESS IS NOT ...

> **Firefighter success is not strictly determined by:**
> - Rank, position, or power,
> - Seniority,
> - Accolades,
> - Time on the job,
> - Salary, or
> - How many fires, rescues, or calls a firefighter has under their belt.

Each of these qualities are noteworthy, but they are not requirements to achieving success. I know firefighters who can check off all of the aforementioned list, but they are not successful because they only care about themselves. They are arrogant, rude, and don't want to help other firefighters succeed. All that they care about is what is best for them, not what is best for everyone else and the fire service.

> "Success is a journey. Make it your own."

At the end of our careers, we will be able to say we gave it our all. We will be able to say we made a lasting impact on our communities, our fellow firefighters, our fire departments, and the fire service as a whole. The time that we spent being firefighters will have positively changed everyone and everything we touched.

HOW TO USE THIS BOOK

Firefighter Success is divided into 20 chapters. Each describes a core principle for success that begins with a C. You may have learned about these principles before, but we will discuss how each specifically applies to you as a firefighter and the fire service. You will see recurring themes throughout the book, and some of the principles share very obvious commonalities—*this is not an accident.*

> "We will leave a legacy worth following."

At the end of each chapter, you will find *Action Steps*, which will give you practical ways to implement *Firefighter Success*' principles. Some chapters will have exercises and questions. To get the most out of our time together, I encourage you to complete them all, giving them serious thought and effort.

I recommend that you initially read *Firefighter Success* straight through, without pausing, to meditate on the details. Then go back through each chapter more methodically—*highlight, make notes, take photos of quotes and passages that impact you, share them with friends and on social media, etc.* Most importantly, apply the principles and Action Steps to your daily life and career.

If you enjoy what you read and want even more great content, head to FirefighterSuccessBook.com. There you can download the free *Special Report: 101 Rules For Firefighter Success* and also learn from successful firefighters and fire service leaders on the *Firefighter Success Podcast*.

Lastly, I want to thank you for your dedication to the fire service. Never stop raising the bar. Never stop learning. Never settle for mediocrity. And don't let anyone stand in your way of achieving success. You are meant for greatness. Stay safe out there.

Jim Moss
Author of *Firefighter Success*

MY BEST SKILL WAS THAT I WAS COACHABLE. I WAS A SPONGE AND AGGRESSIVE TO LEARN.

— MICHAEL JORDAN —

FirefighterSuccessBook.com

CHAPTER 1
COACHABLE

Michael Jordan is arguably the most talented basketball player of all time. With six NBA championships, five Most Valuable Player awards, and over 32,000 points scored, his impact on the sport of basketball is undeniable. He definitely had talent, but that isn't what determined his overwhelming success. Jordan admits that his best skill was being coachable. He was willing and able to receive criticism and guidance—which he used to learn, change, improve, and eventually succeed.

> "The greats never do it on their own."

The greats never do it on their own. If we want to become successful firefighters, we must be willing to accept the fact that we don't know it all, and we will never know it all. No matter how much we train, how many classes we take, or how long we have been in the fire service—*we will never know every single aspect of this job.*

Successful firefighters are coachable. Whether we have three days or three decades on the job, we are willing to receive criticism, process it, and use it to improve. We look to be taught by someone who has walked before us. We know that it is okay to not have all the answers. We know that it is okay to admit, *"I don't know, but I will find out."*

We embody what it means to be humble, because we know that humility is the beginning of our success.

CHAPTER 1

5 KEY ELEMENTS OF HUMILITY

> "There is always someone who is better than you ... at something."
> -Jocko Willink

Before we discuss how to be coachable or anything else contained in this book, we must realize that humility is the foundation. It is the launching point for everything else that is connected to our success. A firefighter without humility will never reach their full potential, nor will they ever really achieve success. Let's discuss five key elements to growing in humility so that we can lay the groundwork to growth and greatness.

1. ADOPT AN "OTHERS FIRST, I AM LAST" MINDSET.

Whether we hold rank or not, successful firefighters are servant leaders. We embody servant leadership and humility by always putting others first. We see the needs of others and strive to meet them on a daily basis.

> "True humility is not thinking less of yourself; it is thinking of yourself less."
> -C.S. Lewis

Isn't that why every firefighter signs up for the job? To serve others? That is why we are called "public servants." We exist to serve. We exist to respond to others in their time of crisis. It is in our DNA. Our careers would be incredibly short-lived if we are dispatched to an alarm, yet say: *"No, we're not going to respond. Have another unit fill in."* That attitude would not go over well.

For our fellow firefighters, a simple way to put them first is by simply showing compassion. On a personal level, if someone on our crew is obviously struggling, let's ask them if we can help. And if we cannot help them, let's find them the help that they need. President Theodore Roosevelt once said, *"People don't care how much you know until they know how much you care."* Let's show our brothers and sisters that we care.

On a lighter note, a practical and simple way we can serve our fellow firefighters is by cooking for them. Perhaps the quickest way to a firefighter's heart is through their stomach. And once everyone sits down to the table to eat, we can be the last to get our food and eat. To paraphrase Simon Sinek: *"Successful firefighters eat last."* When we put others and their needs first, it demonstrates humility.

2. NO TASK IS EVER BELOW US.

Regardless of our position or power, we must know that we are never too important to complete the smallest, mundane task. Whether it is sweeping a staircase, folding towels, or cleaning the fire truck—we will take *pride and ownership* in every aspect of the job. If it needs to be done, we will do it, and we will do it with a positive attitude.

> **"No matter our rank, seniority or time on the job—no task is ever below us."**

Entitlement isn't part of a successful firefighter's vocabulary. Unfortunately, there are some senior firefighters who believe they have earned the right to "do nothing." They believe the fire service owes them everything. But their sense of entitlement and puffed-up egos are roadblocks to their success. On the contrary, senior firefighters have an obligation to do more. The successful veteran firefighter will own the great responsibilities they have to lead by example and to mentor younger firefighters.

When I was a new firefighter, I once found my captain washing the fire truck by himself in the middle of the afternoon. He could have very easily delegated the task to our crew (and especially to me, since I was the least senior member). But he chose to take care of it himself, because in his eyes, *"it just needed to get done."* He saw a need and he filled it. That's what all successful firefighters do when they encounter a problem—*they solve it!* That same captain is now the assistant chief of the fire department, which is no coincidence.

Let's flip the script. Maybe we are the rookie, a private, a driver/engineer, or the like. We will still take ownership of everything that is in our power. Nothing

> **"See a need? Fill a need."**

makes a fire officer happier than when one of their firefighters comes to them and says, *"I found a problem, and I fixed it myself."* As Chief Frank Viscuso has shared in his book *Step Up and Lead*: *"Never walk past a problem you can solve."* See another crew's dirty cup that was left on the kitchen counter? *Solve it.* Find a dirty and rusty hand tool on your fire truck? *Solve it.* Is a piece of the fire truck's equipment broken? *Solve it.*

One of my mentors taught me that there is never a reason to be bored at the fire station. There will always be something for us to learn, train on, fix, clean or do. If you're bored at the firehouse, you're doing it wrong. Consider these routine fire station duties that every firefighter can do:

- Clean trucks, tools, kitchens, bathrooms, etc.,
- Inspect and maintain equipment on our trucks,
- Brew coffee,
- Start and empty the dishwasher,
- Stock toilet paper and towels,
- Clean the apparatus bay floor,
- And many more ...

3. ADMIT AND OWN MISTAKES.

Everyone makes mistakes. Everyone falls short. Everyone fails from time to time.

> "Making mistakes is better than faking perfection."
> -Anonymous

Are we humble enough to admit to our mistakes? Are we prepared to take full ownership of them? Are we willing to learn from our mistakes so that we never repeat the same one? Are we willing to teach others the lessons that we have learned so that they don't suffer the same fate? *Most importantly: If we fail, do we have the courage to stand back up and try again?*

Successful firefighters know that making mistakes is part of the journey. Consider how many times Albert Einstein failed, yet he is known as one of the most brilliant people of his time. Take it from him: *"Anyone who has never made a mistake has never tried anything new."* Thomas Edison is known as one of the greatest inventors in history, yet he gives this advice on failure: *"I have not failed. I've just found 10,000 ways that won't work."*

HUMILITY IN LEADERSHIP

Successful leaders are humble. They aren't scared to take ownership of their errors because they will use them as learning experiences. When it comes to humility, two of the most powerful tools that a leader can use are *honesty* and *transparency*—especially with their mistakes.

> "When we FAIL, sometimes that is just our First Attempt In Learning."

Let me share a story about Jake. Jake was my lieutenant when I first got promoted to the position of captain. He may have only had five years as a firefighter before getting promoted, but he was smart, aggressive, and, most importantly, *he was humble and coachable.* When it was his turn to be on the fire truck with me, I made sure that he was always riding the front seat, responsible for all the calls we responded to.

One night we were dispatched to a first-due working commercial fire at a massage parlor. If you have ever been in one, they are divided up into many rooms so they can host multiple clients. This particular building was 30' wide and 60' deep, and chopped up into approximately 15 rooms.

As we pulled up to the scene, we saw heavy smoke issuing from the right and rear side of the building ("Delta" and "Charlie" sides, for all you firefighters). Jake had been in this building before, and he was very confident about its layout. Since I was riding as the backstep firefighter, he told me to pull an 1.75" attack line to the front door. Second guessing him on his decision, I suggested that we pull a 2.5" attack line to the rear, since that is where most of the smoke was coming from. He respectfully disagreed, and I let him have the autonomy to "make the call." After all, he was the acting officer in charge of the scene.

After we forced the front door, we were met with heavy smoke in the front foyer. Visibility was terrible and there was a decent amount of heat, so I made sure our crew stayed close together. As extra insurance, I told Jake to request that the incident commander position an additional crew with an attack line to the rear of the structure. We attempted to make the push and locate the seat of the fire for about 5 minutes, using our thermal imaging camera and hooks to pop ceiling tiles. After going into several rooms and being halfway into the building without locating the fire, I finally asked Jake: *"Hey L.T., how about we reposition to the rear of the building since we can't find the fire?"* He agreed, and we exited the building.

As we were repositioning our attack line to the rear, we found two crews that were already extinguishing the main body of the fire. They found it right away by going to the rear of the building and making entry through glass sliding doors. We ended up assisting them by performing overhaul and extinguishing hot spots.

As a first due crew, it's never a good feeling to know that someone else found and put out "your" fire. As a first due crew, everyone expects you to make all the right decisions, in only a matter of seconds. But guess what—*that doesn't always happen*. We are human, and we make mistakes.

What went wrong?

Jake did a 360 walk-around of the structure like he was supposed to. He didn't see any obvious fire or smoke from the immediate rear of the building. Judging the conditions he saw when we pulled up, and based on his previous knowledge of the building's layout, he was almost positive we would be able to go in through the front door, make a right turn, and the fire should be right there … *but it wasn't.* Here's what Jake overlooked during his 360: At the rear of the building, there was a 6-foot tall wooden privacy fence that blocked his view of the rear glass sliding doors on the Charlie/Delta corner. He chose to walk around the privacy fence instead of walking *through the gate* (immediately adjacent to the building). If he had gone through the fence's gate, he would have seen the fire's orange glow through the glass sliding doors, and he would have placed the initial attack line at the rear of the structure. Our crew would have immediately found the fire and extinguished it without delay.

Here's the most important part of this story: At our shift's official "After Action Review," Jake had the humility, honesty, and courage to stand up in front of the entire room, admit that he made a mistake, and share what he would do differently next time. That's no simple task, especially as a new lieutenant in a room with 25 other firefighters with strong personalities and opinions.

Jake's outward display of humility demonstrated to other firefighters, fire officers, and chiefs that he had made an error in judgement, but he learned from it. Jake wasn't the only person who learned from the experience. Because of his humility and willingness to admit his mistake, everyone who heard his story gained the knowledge and experience to avoid repeating it in the future.

Moral of the story: Let's have the humility to not only admit to our mistakes and learn from them but also the courage to share with others the lessons we have learned.

4. TALK LESS, LISTEN MORE.

> "A fool speaks, the wise man listens."
> -Ancient Proverb

When we are new firefighters or new to a fire department, almost all of us have been offered the same advice: *"You have two ears and one mouth, so that you can listen twice as much as you speak"* (Epictetus). Some may turn their noses up at this advice and call it old-fashioned, but there is great truth to it. But this advice isn't only applicable to new firefighters—every firefighter could talk less, and listen more.

If you have been in the fire service for any amount of time, consider how many times someone has stuck their foot in their mouth and made a fool of themselves because they speak without thinking *(perhaps we have done it many times).*

> "A firefighter with humility will be *quick to listen, slow to speak, and slow to become angry."*
> -Book of James
> *(paraphrased)*

Contrary to popular belief, it is okay to "engage our filters" and to not say every thought that comes to mind. I have witnessed rookies, senior firefighters, company officers, and even chiefs make complete fools of themselves by blurting out obscene and offensive comments—all because they did not think before they spoke. We all have done it, myself included, and when it happens we immediately regret what we said. We wish we could take it back and that no one would have ever heard it.

Imagine how many arguments could be avoided if we simply thought and engaged our filters before we spoke. We can practice the THINK principle by asking: "Is what we are about to say:

| True? | Helpful? | Inspiring? | Necessary? | Kind? |

5. RESPECT BOUNDARIES AND EMBRACE DISCIPLINE.

For those of us who have children, we understand the importance of boundaries, enforcing them, and administering discipline. Without them, a child will grow up to face a hard life in the real world. The same goes for firefighters: We must have the humility to respect the boundaries and expectations placed in front of us (i.e. "Rules and Regulations," "Standard Operating Procedures," etc.).

When we don't live up to expectations, or we don't follow the rules, corrective action needs to take place. A humble and coachable firefighter will view discipline as guidance and a means to correct the undesired behavior, *not as punishment*.

As a firefighter, I have been on both sides of discipline: receiving it and administering it (as a company officer). When I was on the receiving end, I accepted it and admitted to my mistakes. With humility, I apologized and pledged to not make the same mistake again.

As a company officer who also administers discipline, I utilize the following process:

1. Allow the accused to give their side of the story first.
2. Inform them of how they broke the rules or didn't meet expectations.
3. Reinforce expectations, being very clear.
4. Communicate what will happen the next time they break the rules.

BE COACHABLE, BE TEACHABLE.

With humility as our foundation, we are ready to discuss what it means to be coachable, why it is important, and how to live it out.

A coachable, successful firefighter is *teachable*. When others give us advice or criticism, we don't roll our eyes and merely shrug it off. We check our ego at the door and we receive it. Not only do we receive it well but we also actively seek out others' guidance on the path to continuous self-improvement.

Successful firefighters also know training is essential to every aspect of the job. Training teaches skills, imparts knowledge, and instills confidence in us. As coachable firefighters, we are always willing to train and we have an unquenchable thirst for learning everything we can about our profession. We invest in our own knowledge, skills, and abilities by seeking out training opportunities in every avenue:

- Reading articles and books
- Building and area familiarization
- Attending conferences
- Hands-on training
- Listening to podcasts
- Analyzing calls and learning from them

In *Chapter 8 - Competent*, we will discuss firefighter training in great detail and provide over 100 hands-on training ideas.

Lastly, as coachable firefighters, we are honest about our strengths and weaknesses. We do not shy away from personal introspection because we know it is critical to our growth. Knowing who we truly are, we work hard to change bad habits and improve our weaknesses.

MENTORSHIP IS ESSENTIAL

We cannot achieve success on our own. We must seek the guidance of someone who is wiser and more experienced. A mentor is integral to our success, because they will:

1. Have the ability to see and point out weaknesses we couldn't see in ourselves.
2. Teach us from their wisdom, experience, and failures.
3. Increase our confidence.
4. Provide encouragement when we fail.
5. Hold us accountable to our goals.
6. Fast-track our success.

Finding a mentor isn't complicated. Ask yourself: *"Who is a leader I know who I aspire to be like?"* Undoubtedly we can identify one to three individuals who would be willing to meet and talk with us on a regular basis. There is no need to have a formal relationship or a written contract. We can simply ask our prospective mentor if they would like to meet for a coffee or a meal. At our meetings, let's be honest about our goals so that they can help us. Also, let the mentor know why we admire them, and ask them if they would be willing to meet on a regular basis to discuss how we could learn from their experience. At no point do we need to use the words "mentor" or "mentorship." Just keep it informal and easygoing.

> **"A mentor is one who knows the way, goes the way, and shows the way."**
> **-John C. Maxwell**
> *(paraphrased)*

WHAT HAPPENS WHEN A FIREFIGHTER IS NOT COACHABLE?

Ego, vanity, selfishness, negativity … these are all symptoms of a firefighter who refuses to be coached. Their "know-it-all" attitude is a roadblock to reaching their full potential, and consequently a barrier to the team's success. An uncoachable firefighter refuses to be humble, refuses to be teachable, and views discipline as punishment.

Every fire department has uncoachable, ego-driven firefighters. Stay away from them at all costs, because their poor attitudes will be a cancer to everyone around them. They will infect others with their negativity in no time.

MAKE THE CHOICE TO BE COACHABLE

Being coachable is a daily attitude. Day in and day out, it is a mindset that must be developed. It is a choice that we as firefighters must make if we truly want to unlock our potential and achieve success. Coachable firefighters are humble, teachable, and are constantly seeking to improve themselves.

> "Stay hungry, stay humble, and stay hopeful."
> -Onyi Anyado

ACTION STEPS

1. Time for some honest introspection. Ask yourself: *Do I need to be more humble? Are there specific areas in my life that I need to put my ego aside?*

2. Write down your strengths and weaknesses. Don't hold back.

3. Don't pretend to have all of the answers. Receive criticism and use it to get better.

4. Find a mentor.

5. Commit to change, improve daily, and most importantly: *take action.*

YOU CAN NEVER LEARN TOO MUCH ABOUT A JOB THAT CAN KILL YOU.

— CAPTAIN TOM BRENNAN, FDNY —

FirefighterSuccessBook.com

CHAPTER 2
CURIOUS

One of the greatest travesties in the fire service is that rookies are told to *"sit down, shut up, and don't ask questions."* Don't get me wrong—I completely understand and value the role of a rookie firefighter. All of us have been or will be "the rookie" at some point in our careers. Some of us have even been the rookie at multiple fire departments.

> "Curiosity is the wick in the candle of learning."
> -William Arthur Ward

It is a given that rookies must earn the respect of their crewmates and prove themselves by taking on extra responsibilities. They must also prove themselves to be reliable at emergency scenes. Rookies have quite the weight to bear during their first couple of years: learning their craft, assimilating themselves to their crew and the fire department's culture, extra station duties, etc. But by proving themselves on a daily basis, they are showing their fellow firefighters they are trustworthy. They are proving they are worthy enough to be part of the team.

As a company officer and instructor, one of my greatest joys is taking on a new firefighter. There are many reasons I enjoy having a rookie on my crew: orienting them to our crew, assimilating them to our fire department, and training them to be part of our team. *What I admire most about taking on a rookie is that they are always curious.* Their willingness to learn is like a breath of fresh air and their passion is contagious to the rest of the crew. As other crew members see the rookie's curiosity and willingness to train, it inspires them to teach the rookie more of the tips and tricks of the trade.

For example, a rookie may have only learned one or two ways to force a door in the fire academy. Their more-experienced crew members will likely show them several more ways and specific techniques for greater mechanical advantage and efficiency. When senior firefighters see a rookie who is curious and "like a sponge," they are more willing to mentor them and take them under their wings.

CURIOSITY STARTS WITH ASKING "WHY?"

> "Whether young or old, rookie or veteran, we must always ask ourselves: *Is there a better way?*"

Contrary to most of the fire service's mentality, I actually want my rookies to ask "*Why?*" I want them to understand *why* we do things the way we do. I know that when they understand the process and the reason behind it, it creates greater buy-in and ownership. Additionally, if they know the reason behind everything they do, they will be able to fully teach the next rookie. Albert Einstein once said, *"Any fool can know. The point is to understand."* In other words: Anyone can know how to do a task, but it is more important to know *why*.

Furthermore, successful firefighters know knowledge is power, and knowledge instills greater confidence and competence. The more we understand, the more effective and safer our team will become.

Being curious and always willing to learn aren't only necessities for the rookie. Every successful firefighter must possess them. If we are curious, it means that we have an open mind, we are pursuing knowledge, and we are teachable (as we discussed in Chapter 1). If we stay curious about learning our entire careers, it means we are always aiming to improve ourselves and the way that we do our job. It means we are always improving our level of professionalism and our service delivery to our public.

> "Any firefighter can be taught *how* to force a door, but it is more important to know *why*. Once we know *the why*, *the how* comes naturally."

SUCCESSFUL FIREFIGHTERS SEEK AND IMPLEMENT THE BEST METHODS TO ACCOMPLISH THE TASK AND THE MISSION. WE WORK SMARTER, NOT HARDER. WE FIND THE MOST EFFICIENT AND MOST EFFECTIVE WAY TO GET THE JOB DONE.

FirefighterSuccessBook.com

CHAPTER 2

INNOVATION AND CREATIVITY

Curiosity breeds innovation. As we will discuss in *Chapter 18 - Change*, the fire service has grown and survived because of our ability to adapt and evolve with change. As innovative firefighters, we look at the status quo from a different and fresh perspective. We ask questions like: *"Is there a better way to advance this hose line?" "Is there a better way to tie this knot?" "Is this Standard Operating Guideline (SOG) still relevant to what we do?"* Some of the best training happens when firefighters get together and think outside of the box.

Our curiosity also begets creativity, and gives us permission to ask, *"What will happen if we try it this way?"* Sometimes new ideas do not work, and yet sometimes they provide us with a better way. The best idea can come from anyone of any rank and time on the job—and it will always come from a curious firefighter who is willing to be creative.

In today's fire service, we face countless obstacles within our own fire departments—limited funding, fewer resources, recruitment and retention issues, generational differences, etc. Curiosity, innovation, and creativity must be fundamental elements of every firefighter's mindset and skill set. We will do the best we can with what we have been given. We will maximize our own potential and our fire department's potential by adapting, overcoming, and eventually succeeding.

A WORD OF ADVICE FOR CURIOUS ROOKIES

Without a doubt, some senior firefighters are aggravated by a new employee asking them *"Why do we do it this way?"* They feel threatened that the curious rookie is second-guessing the way things have been done for decades. They may also feel embarrassed because it is quite possible that they do not know the answer. (Queue the parent-type responses of *"Because I said so," "Because that's how we do it,"* and *"Stop asking questions, kid."*)

There is a proper way, time, and place for the rookie to ask *why*. First and foremost, the rookie must ask in a manner that is curious and not confrontational. They must never come off as challenging or standoff-ish. Instead, they should use questions like: *"Could you explain that a little more?"* Or: *"I like how you did that. Could you show it to me again? Why did you do it that way?"*

> **"A rookie must first build respect at the individual level, then at the crew level, then with their shift, and finally with the entire fire department."**

As it pertains to time and place, it is best for a rookie to ask questions during a one-on-one training session. For example, if they are training on forcible entry with only their officer, there is a higher comfort level and more freedom to ask virtually whatever they want. However, if they are training with their entire shift in a bigger group setting, they should wait to ask a lot of questions until the training is done and they are back at the station with their officer or their crew.

In my experience, when a rookie asks too many questions in a large group setting, it reflects poorly on them. It may also reflect poorly on the rookie's crew. Doing so leaves a bad first impression on veteran firefighters, which typically leads to a negative opinion and gossip. You may hear things like, *"Man, did you hear how many questions he asked?"* Or: *"I just wanted to get out of there, but she just kept asking questions. She just wouldn't shut up."* All of this may sound preposterous, but it is true.

Rookies must also have tact in how they express themselves, understanding that it is crucial to build rapport and respect with the other firefighters at their new fire department. Treading lightly, they must avoid saying, *"Well, that's not how we did it at my old fire department."* That phrase is an automatic respect-killer. Rookies: Do not expect other firefighters to teach you anything if that is your attitude.

A second piece of advice for rookies: Avoid sharing war stories from your previous fire department. Too many rookies have made this mistake and then were labeled as the guy or gal who constantly brags about their previous fire department. Some veteran firefighters have even said, *"If you like talking about your previous fire department so much, why don't you just go back?"* That is a phrase that no rookie wants to hear.

> **"There is a proper way for a rookie to ask why. It must always be done with tact, humility, and respect."**

LEARNING IS A LIFELONG JOURNEY

Let's be honest, it is much easier to be curious and teachable at the beginning of our careers. We are excited about being a new firefighter, and we are passionate about every single aspect of the job. But if we want to get the most out of our time in the fire service, we will be curious learners, day-in and day-out, year after year. Not only that, our lifelong curiosity will allow us to give back to the job we love.

CHAPTER 2
ACTION STEPS

1. Are you a rookie? Be curious, ask why, and learn something new every time you are at the firehouse.

2. Are you a veteran firefighter? Be curious, ask why, and learn something new every time you are at the firehouse.

3. Company officers, training officers, and instructors: Encourage your members to ask questions. Encourage them to be creative, innovative, and curious.

CURIOSITY, INNOVATION, AND CREATIVITY MUST BE FUNDAMENTAL ELEMENTS OF EVERY FIREFIGHTER'S MINDSET AND SKILL SET. WE WILL DO THE BEST WE CAN WITH WHAT WE HAVE BEEN GIVEN. WE WILL MAXIMIZE OUR OWN POTENTIAL AND OUR FIRE DEPARTMENT'S POTENTIAL BY ADAPTING, OVERCOMING, AND EVENTUALLY SUCCEEDING.

FirefighterSuccessBook.com

IN ORDER TO BECOME THE 1%, WE MUST DO WHAT THE OTHER 99% WON'T.

FirefighterSuccessBook.com

CHAPTER 3
CHAMPION MINDSET

Mindset is everything. It is what separates the best from the rest.

Everything begins in the mind. Before we can achieve anything, we must believe we can achieve everything. If we are to achieve success and attain excellence, it all starts with a champion mindset.

Our minds are like muscles—the more we exercise them, the stronger they get. The more we flex them, the more they expand and the more we grow. As Dan Kerrigan and I have shared with *Firefighter Functional Fitness*, fitness is 90% mental, and 10% physical. In other words, if we cannot first convince our minds to convince our bodies to train, we will never make progress. The power of the mind simply cannot be overstated.

> "When we lead our minds, we are able to lead anything."

We will discuss three essential elements of the champion mindset: *a positive mindset, an unstoppable mindset, and a competitive mindset.*

CHAPTER 3

POSITIVE MINDSET

Too many people focus on what is going wrong in the world. They are great at complaining about the problems in their lives, but they conveniently never have enough time or energy to come up with any solutions. They end up focusing all of their thoughts on what could be better, instead of taking action to make their lives better. They end up making themselves (and others around them) absolutely miserable.

In the same vein, pessimistic people tend to lack confidence and drive. They are so crippled by the thought of failure that they sit on the sidelines, too scared to even try. I have known too many firefighters who have decided to not go for a promotion because they claim they are poor test takers, they don't believe they are smart enough to beat out their competition, and they ultimately fear failure. Unfortunately, their own mindset defeated them long before they even made an effort.

> "The pessimist complains about the wind.
> The optimist expects it to change.
> The realist adjusts the sails."
> - William Arthur Ward -

Let's do away with all negativity.

Let's extinguish words like *can't* and *impossible* from our vocabulary.

Let's see opportunity in every difficulty by framing our entire lives with a mindset of positivity. As positive firefighters, we won't focus on what could go wrong. Instead we envision what will happen when we succeed. With a positive outlook, we see things differently. The more we improve our mindset, the more things around us improve.

Do we have poor leadership in our firehouse or fire department? *Let's worry about leading ourselves and also leading by the example that we want others to emulate.*

> "The only way to truly guarantee failure is to never try."

Do we have a crew of complainers? *Let's focus on the positive of each firefighter and see if perhaps they have deeper underlying issues.*

Does our fire department have substandard equipment? *Let's make the most of what we have for now, and let's also apply for grant funding.*

When we adopt a positive mindset, it shapes our daily attitude. Our attitude is a choice—a choice that we make every single day when we wake up and get out of bed. Our attitude determines our thoughts, our thoughts determine our behavior, our behavior determines our actions, and our actions determine the outcome. So let's choose to start every single day with great attitudes.

Every day will present itself with its own unique struggles. We can choose to surrender to adversity, or we can succeed with a champion mindset.

Let's be positive. Let's take action. Let's believe in our success. *It will come.*

> "We will be part of the solution, and not part of the problem."

The last element of the positive mindset is adopting an attitude of gratitude. Whether we have a little or a lot, whether we have poor or good leadership, whether we have a bad or good crew—let's be thankful for what we have. The more we say "thank you," the less opportunity we have to complain. Germany Kent said it best: *"It's a funny thing about life, once we begin to take note of the things we are grateful for, we begin to lose sight of the things that we lack."*

And if we are the leaders within our fire departments, the words "thank you" must roll off our tongues every chance we get. Even if our members are "just doing their jobs," giving gratitude goes a long way. Every single firefighter wants to know their efforts are recognized and that they are appreciated. Did one of our firefighters cook dinner for the crew? Say *"thank you."* Did one of our firefighters clean the firehouse? Say *"thank you."* Did our crew do a good job on a call? Say *"thank you."* These words will never be wasted.

UNSTOPPABLE MINDSET

Elie Wiesel was fifteen years old when he was brought to the Nazi concentration camp Auschwitz during the Holocaust. While he endured countless acts of abuse, both of his parents died at the hands of their Nazi captors. He survived and was eventually liberated, but he had to start his new life on his own as an orphan. Even though he suffered unspeakable atrocities, his *unstoppable mindset* is what helped him to succeed in life. Turning tragedy into triumph, he became a best-selling author and human rights advocate around the world, using his voice for those who had no voice. Wiesel could have given up in life, letting the Nazis win. But instead he chose to fight.[2]

> "If something stands between you and your success, move it. Never be denied."
> -Dwayne Johnson

Do you have the *unstoppable mindset*?

---

2 "Elie Wiesel, Holocaust survivor and best-selling author, is born." *History*, June 30, 2011, www.history.com/this-day-in-history/elie-wiesel-holocaust-survivor-and-best-selling-author-is-born.

AS SUCCESSFUL FIREFIGHTERS, WE KNOW THAT NO ONE WILL HOLD US BACK FROM ACHIEVING OUR GOALS AND NO ONE WILL STOP US FROM SUCCEEDING — THIS IS THE UNSTOPPABLE MINDSET.
EVEN IN THE FACE OF OVERWHELMING ODDS, OUR DETERMINATION, PERSISTENCE, AND SELF-DISCIPLINE WILL ENSURE THAT WE BREAK THROUGH ANY OBSTACLE THAT STANDS IN OUR WAY.
WE WILL NOT ACCEPT FAILURE, BECAUSE WE KNOW THAT ADVERSITY BUILDS TRUE CHAMPIONS.

FirefighterSuccessBook.com

Others will try to place limits on us. They will try to hold us back. We won't preoccupy ourselves with proving them wrong; rather, we will only seek to prove ourselves right. We will use the criticism of mediocre firefighters to fuel our success. With their bricks of condemnation, we will build a foundation for excellence.

Conceive. Believe. Achieve. If our minds conceive a goal, we will unwaveringly believe in ourselves and we will eventually *achieve* it.

The question is not: *"Will we succeed?"* The question is *"How long will it take us to succeed?"*

A fundamental element of the unstoppable mindset is our unwavering commitment to continuous growth. Every day, we seek to realize and expand our potential. We aren't content with sitting on the sidelines of life—it is our goal to thrive and be successful.

> "Do it on your own. No one needs to give you permission to be great."
> -Capt. Mark vonAppen

While other firefighters are content with being average, we choose to do something different every day to get out of our comfort zones. We are willing to sacrifice while others want to be comfortable. *That is why we learn every day, we read every day, we train every day.* Sometimes growth comes quickly, and sometimes it is a long road. Sometimes we succeed right away, and sometimes we have setbacks. It doesn't matter how slow we go. As long as we are moving, we are making progress. Above all, our unstoppable mindset motivates us to be better than we were yesterday, and that is what continuous growth is all about.

COMPETITIVE MINDSET

We as successful firefighters use competition as part of our champion mindset. We are always in competition with ourselves and having such a mindset means that we are never satisfied with mediocrity. As Jean Giraudoux once put it, *"Only the mediocre are always at their best."*

We are always hungry to improve: physically, mentally, tactically, emotionally, etc. We are our biggest critics, knowing that there is always room for improvement. Our insatiable drive is what feeds us to become the best versions of ourselves that we can possibly be. We know that we will never achieve perfection, but in striving for it we will maximize our success.

> "You may be bigger, faster, stronger, and smarter than me, but you won't outwork me."

> "When it comes to the fireground, no one will work harder than us. When it comes to training, no one will train more than us. When it is time to prepare for a promotion, no one will study harder than us."

On the other hand, being in competition with others can become a double-edged sword, especially when we start to *compare* ourselves to others. We must know the difference between healthy competition and comparison. Comparison may negatively impact us because we can lose focus of our goals if we end up idolizing someone else's goals, progress, and success.

For example, it may be another firefighter's goal to eventually become chief of the fire department, but that shouldn't necessarily be our goal. Or someone who was in our fire academy class may now be at the chief level, and we are still at the firefighter level. Such unnecessary comparisons may cause us to have unrealistic expectations for ourselves. A brand-new rookie cannot possibly compare themselves to a chief who has been on the job for 30 years—their experience, knowledge, and skill levels are in different leagues.

On the other hand, *healthy competition* with those who are at our level or just ahead of us will push us to become better than we ever thought possible. When we train with others, especially for workouts and hands-on training, our competitive drive kicks in and we work harder. We don't want to be seen as the weak link, so we end up giving 110% effort. This form of positive peer pressure is known as the Köhler effect.

Additionally, we can inspire others to up their game with healthy competition—taking their knowledge and skills to the next level. Larry Bird and Magic Johnson, Tom Brady and Peyton Manning, Muhammad Ali and Joe Frazier—all of these legendary athletes admit that competition (and a bit of rivalry) played a large role in their success.

> "Our crew is only as strong as its weakest link. We will not be the weak link."

It may sound odd, but firefighters can use healthy competition to study with the other firefighters who are going through the same promotional process. We will all have specific concepts that others haven't covered, and others will have material that we haven't studied yet. Some may argue that this is "helping the competition," but what is really doing is improving everyone's abilities. As the old adage goes: *A rising tide raises all boats*. So, let's not be scared of a little competition, as long as it is the healthy variety.

LEAD YOUR MIND, LEAD YOURSELF

If we have power over our minds, then we have power over everything else: our attitudes, actions, and eventually our success. Possessing the champion mindset comes back to being positive, being unstoppable, and being competitive. Day in and day out, our mindset is our only limitation. When we lead our minds, we can then lead ourselves in achieving excellence.

ACTION STEPS

1. What are some negative things in your life right now? How are you reacting to them? Proceed with a strong, positive mindset.

2. Make it a point to say "thank you" more this week (e.g. at work, at home, and out in public). Take note of how having an attitude of gratitude makes you feel.

3. With your goals, what are the specific obstacles that confront you? Be honest. Are there legitimate reasons that are inhibiting your progress, or are you just making excuses? Adopt the unstoppable mindset and proceed with such fierce determination that no one will stop you from achieving your goals.

4. Identify two to three firefighters who will push you to become better. Engage in healthy competition with them on a regular basis (e.g. training, workouts, studying for a promotion, etc.).

TRAINING PRODUCES THE SKILLS.
CIRCUMSTANCE PROVIDES THE OPPORTUNITY.
EXPERIENCE INSTILLS TRUE KNOWLEDGE.

FirefighterSuccessBook.com

CHAPTER 4
CONFIDENT

Whether we are the rookie, senior firefighter, driver, or company officer, we will be confident in our position. We must have confidence in ourselves, specifically our knowledge, skills, and abilities. We must also have the confidence to allow ourselves to make mistakes and then the confidence to admit to them and learn from them.

In the simplest of terms, confidence is believing in ourselves, being self-assured, and believing with certainty that we will achieve our goals. It requires boldness, faith, and conviction. Our self-confidence doesn't waver in the face of adversity. As confident firefighters, we do not apologize for who we are, what we believe, or our passion for the job.

> "It's not the size of the dog in the fight, it's the size of the fight in the dog."
> -Mark Twain

CONFIDENCE VS. EGO

We cannot allow ourselves to confuse confidence with ego or arrogance—both of which are barriers to firefighter success. Not only will these build a terrible reputation, they will also destroy our credibility as firefighters. They will be roadblocks to unlocking our full potential. Remember: *Humility is the foundation to everything tied to our success.*

Let's examine *confidence* and *ego* by comparing the following statements. Which do you identify with?

CONFIDENCE	EGO
"Do you mind giving me a hand? I could use some help."	"I don't need your help. I can do it on my own."
"That firefighter is really smart. I could learn from their knowledge and experience."	"They can't teach me anything. They're just a rookie."
"I may not know the answer, but I will look into it for you."	"I know everything. I know more than everyone here."
"Let the best idea win." -Chief Frank Viscuso	"My way is always the best way."
"I trust them to get the job done."	"I will micromanage everything they do."
"I will prove myself every day."	"I've been here for 20 years; I *deserve* this promotion."
"They look like they could use some help. I'm going to ask them."	"They will figure it out on their own."

> "Ego is the barrier to success."

WHY CONFIDENCE IS IMPORTANT

We work in a profession that requires quick decisions that mean life or death—both for ourselves and for the citizens we serve.

For example, when we arrive on the scene of a working house fire in the middle of the night with cars parked outside, we must make critical decisions:

1. Lay a supply line from the hydrant first?
2. Perform fire attack first?
3. Perform search/rescue or vent-enter-isolate-search first?
4. Which attack line to deploy?
5. Where to place the attack line for maximal impact?

This list of questions and decisions could go on and on, but nothing will ever be accomplished if we lack the confidence to make a decision.

Consider the citizens we serve. In times of desperate need, they expect firefighters who are 100% confident in themselves and what needs to be done. They expect a firefighter who is all-in, aggressive, and ready to save them. Make no mistake about it, they can tell when a firefighter is unsure or insecure. But we will be firefighters who are confident to get the job done.

When I was first promoted to the position of lieutenant, I had limited experience. My battalion chief told me: *"Other new officers in your position might be scared to make a hard decision on the fireground or at the firehouse. I know that you will make a decision because you are confident in yourself. Trust me, you will make mistakes when making decisions, and that's okay. We can deal with those later. What I can't tolerate is someone who cannot make a decision."*

> **"Be the firefighter who is confident to get the job done."**

What was my battalion chief telling me? *Be bold. Be confident. Be sure of yourself. Make the call and own it.*

> **"The worst decision is indecision."**
> -Ryan Harwood

Chief officers want company officers and firefighters who are confident, decisive, and willing to make mistakes. Company officers expect the same from their firefighters. And if you are a company officer, have the confidence in your crew to get the job done. Give them the autonomy to make decisions and complete the tasks the way that they see fit. When we micromanage our people, we are expressing a lack of confidence in them. They will undoubtedly reciprocate a lack of confidence in us as their company officers.

10 WAYS TO BE MORE CONFIDENT

1. BE OPTIMISTIC.

Having a positive mindset is crucial to our daily attitude and confidence. Let's remember that our attitude is contagious—let's have one that is worth catching.

2. SMILE AND LOOK OTHERS IN THE EYE.

The majority of effective communication isn't the words that we say. Rather, it is our body language and how we project ourselves. Let's project confidence with a smile, eye contact, and a handshake.

3. VIEW FAILURE AS OPPORTUNITY.

Have the confidence to fail! How we view failure will determine how far we ultimately advance. Remember: When we FAIL it is our First Attempt In Learning. Always, always, always … Get back up and try again.

4. MAKE OUR WORDS MATTER.

"Mean what you say and say what you mean." If we say something, say it with confidence ... and then stand behind it! But remember, we must think before we speak. If we say that we are going to do something, let's do it. Words mean absolutely nothing if we do not back it up with action.

5. DEVELOP OUR STRENGTHS.

We have strengths for a reason. We must use and develop them to our advantage. What are we passionate about? *Leadership? Engine company operations? Truck company operations? Rescue techniques? Fitness? EMS?* Let's grow in our passions and fuel them so we can grow in confidence.

6. SET ATTAINABLE GOALS.

When we set goals, let's make sure that they are attainable. Use the SMART goal-setting process: specific, measureable, attainable, relevant, and timebound. By completing smaller goals, we will grow in confidence and be able to establish bigger goals—*and achieve them!*

7. TRAIN.

Train, train, train ... and then train some more. Building knowledge and skills through repetition is a great method to grow in confidence. Perform at least 30 minutes of hands-on training per shift (e.g. donning PPE, masking up quickly, rescue breathing techniques, ladders, attack line deployment, ropes and knots, etc.). Refer to *Chapter 8 - Competent* for hands-on training ideas.

8. "DRESS FOR SUCCESS."

The way we project ourselves with our physical appearance says a lot about our self-confidence. The pride we take with our uniform, hair, and physique is communicated to our citizens and fellow firefighters. Who exudes confdence: *An overweight firefighter whose shirt is untucked and hair is a mess? Or a physically fit firefighter who is well-groomed?* Both firefighters may be able to do the job, but their appearances communicate their confidence and professionalism.

9. USE SELF-AFFIRMATION.

What does a weightlifter do before they attempt a personal record? They get psyched up by telling themselves that they can do it. They boost their self-confidence, heighten their mental acuity, and physically prepare their body to go where it has never been before. Use daily affirmations to build the unstoppable mindset: *"I will do it." "No one will hold me back." "I am my only limitation."*

10. GET COMFORTABLE WITH BEING UNCOMFORTABLE.

It's easy for us to stay in our comfort zone. But growth never happens in the comfort zone. Let's challenge ourselves on a daily basis to something new and different that will make us smarter, stronger, more resilient, and more self-confident.

> "Do something every day that you don't want to do; this is the golden rule for acquiring the habit of doing your duty without pain."
> - Mark Twain -

LIVE LIKE LINCOLN

Abraham Lincoln is regarded by many as the greatest president in United States history. Throughout his career, he faced numerous failures and setbacks, including election defeats, losing jobs, and nervous breakdowns. In the long-term, confidence and persistence are what helped him to not only get elected in 1860 to President, but to also be one of the most successful in all of American history.

If we want to achieve success, we will move forward with confidence, simultaneously maintaining a foundation of humility.

ACTION STEPS

1. Set a short-term goal and a long-term goal. Use the SMART method.

2. Identify a weak area in your firefighter skill-set (e.g. knots, ladders, etc.). Make the commitment to drill on that topic for the next month.

3. Do one thing this week that purposely pushes you out of your comfort zone.

4. Take better care of yourself physically. Eat healthier, exercise more, and get more sleep. Investing in your physical well-being will boost your confidence.

MOST PEOPLE FAIL, NOT BECAUSE OF LACK OF DESIRE, BUT, BECAUSE OF LACK OF COMMITMENT.

- VINCE LOMBARDI -

FirefighterSuccessBook.com

CHAPTER 5
COMMITTED

> "Give it everything you've got. Don't just make a living. Make a life. Don't just earn a paycheck. Go after the passions God has put in your heart. Halfway is no way to live; you've got to go all in."
> - Mark Batterson, *Play the Man* -

Whether we are the rookie fresh out of the fire academy or we are the most senior firefighter, we must be committed to our craft. We must invest our time, energy, and passion into a career that is greater than us.

Being a firefighter isn't just a job where we can show up, clock in, and clock out—it is who we are. As David J. Soler shares in *Firefighter Preplan*, *"For some it's a job. For us it's a calling."*

We are *called* to put the needs of others before our own on a daily basis.

We are *called* to sacrifice our comfort, our bodies, our health, our safety, and even our emotional health to serve complete strangers.

We are *called* to be at our best when others are at their worst.

6 VIRTUES OF COMMITMENT

Dedication, honor, duty, loyalty, pride, and passion. Combine all of these virtues into a singular, yet comprehensive idea of what every successful firefighter must possess: commitment.

1. DEDICATION

The committed firefighter devotes their time, skills, knowledge, abilities, and efforts to serve others. When others call, we respond to help—not just some of the time and not just when we feel like it *but all of the time*. Career firefighters will do this while working long shifts, anywhere from 12 hours to more than 72 hours at a time. Volunteer firefighters answer the call from home, work, and even play—all without monetary compensation. Volunteers truly exemplify what it means to be dedicated and committed.

> "There's nothing stronger than the heart of a volunteer."
> -Jimmy Doolittle

As committed firefighters, our dedication goes above and beyond just learning the craft and responding to alarms. We seek additional opportunities to serve our citizens by checking their homes' smoke and carbon monoxide detectors, cutting their grass, shoveling their snow, and taking out the trash when needed. The undedicated firefighter will say, *"That's not my problem,"* while the dedicated firefighter will always go the extra mile to serve others. The next time we are at a citizen's house for a routine call, let's find a way to go above and beyond to serve them.

> "You're either in or you're out. Apathy and mediocrity have no place in the fire service."

2. HONOR

As committed firefighters, we will have great respect for the position we hold. We will carry ourselves with integrity, honesty, and have high moral and ethical standards. Whether or not we are a paid firefighter, we will always be professional in how we act and behave. We will be an example of honor—both in the firehouse and outside of the firehouse. We will choose not to say or do anything that would disrespect the fire service, our fire department, our fellow firefighters, or our citizens.

> "Integrity is doing the right thing, even when no one is watching."
> -C.S. Lewis

As committed firefighters, we will also honor the traditions of the past, knowing how the profession has evolved. We take time to learn fire service history—its beginnings, major events, and how it has changed. We will honor the sacrifice of fallen firefighters by reviewing how they died in the line of duty. From lessons learned and taking our training seriously, we will honor fallen firefighters by not repeating the same errors which may have occurred.

> "If we do not honor our past, we lose our future. If we destroy our roots, we cannot grow."
> -Friedensreich Hundertwasser

3. DUTY

We always answer the call of duty. Not only do we respond to every alarm we are called to but we also own our primary responsibility to serve others. As committed firefighters, we know we have an obligation to be the best that we can be—not for our own virtue, but for the benefit of serving our community.

A firefighter who understands the magnitude of their responsibilities carries the utmost trust from their citizens. Regardless of the type of emergency at hand, John Q. Public has 100% faith that his firefighters will take care of his problem. It could be the most routine and mundane call, or it could be an extremely dangerous situation—*the committed firefighter will respond with duty and commitment.*

4. LOYALTY

A committed firefighter is devoted to the service of others. We are loyal to the brotherhood and sisterhood of our fellow firefighters. We believe in camaraderie and community, and we live them out through our actions. When we see someone is in need (financially, emotionally, physically, etc.), we continue to answer the call by meeting those needs.

There are so many ways to demonstrate loyalty to our team, but the key is to be considerate and helpful. It may sound funny, but one of the simplest ways is to fold a crew member's laundry when it is finished drying. When it is time to transfer our clothes from the washer to the dryer and find that their clothes are dry, it would be very easy to simply take theirs out and just throw it on top of the dryer. However, a loyal firefighter will take the time to fold their fellow firefighter's clothes. This is such a simple act of service that says, *"I've got your back."*

As loyal firefighters, we must avoid gossip—not only listening to it but also speaking it. Whether true or untrue, gossip is any negative speech that is said about another person. The worst kind of gossip is that which intentionally spreads false rumors about

> "Great minds discuss ideas. Average minds discuss events. Small minds discuss people."
> -Eleanor Roosevelt

someone else. Anyone who has been in the fire service for any amount of time knows gossip is a real and significant issue. The next time a fellow firefighter starts to gossip about someone else, let's be the ones who don't participate. Let's be the firefighters who lead by the example of loyalty we want others to follow. Remember, we can use the THINK principle of speaking. *Is it True? Helpful? Inspiring? Necessary? Kind?*

5. PRIDE

A committed firefighter takes pride in every aspect of the job. We take pride in ourselves: *our appearance, our fitness, and our knowledge, skills, and abilities.* We take pride in our uniform, personal protective equipment, and especially our apparatus and all equipment. We inspect everything when we arrive at the fire station to ensure it is 100% operational. We also know where every single piece of equipment is located on our trucks. If something is broken, we attempt to fix it, and if we can't, we immediately report it to someone who can. We maintain our tools by cleaning them regularly. Consider a specific day of the week for regular cleaning and maintenance of our truck's hand tools (e.g. "Tool Pride Tuesday").

> "Dirty tools are not a sign of a busy company. Dirty tools are the sign of a lazy company with no pride."
> -Lt. Doug Rohn
> City of Madison Fire Department

As committed firefighters, we take pride in our stations by cleaning and fixing whatever needs attention. We make sure that our stations' flags are clean, undamaged, at the correct height, and illuminated at night. We never walk past a mess, dirty dish, or any type of problem and say, "That's not mine. I'll leave it for the next guy." Instead we take the time required to clean the mess or pick up the dish.

If the microwave needs cleaning, we do it. If the paper towels or toilet paper need to be replaced, we do it. If the station towels need to be washed and folded, we do it. We take complete ownership of everything within our power, and we do so with the utmost pride.

> "The pride a firefighter takes in doing the small things says a lot about how they will perform on the fireground."

6. PASSION

Of all the virtues we have discussed, a firefighter's passion for the job is the quintessential expression of commitment. A firefighter with passion is enthusiastic about every aspect of the job—serving, training, learning, and even cleaning. It's easy to come to the firehouse when we are passionate about the job, because *"passion equals purpose,"* (Mark Bryant, L.A. County Fire). We know our purpose is to serve others the best we can, and this purpose fuels us on a daily basis.

> **"The biggest fire can start from just one spark."**

Our passion fuels us to work hard at everything we do, and we demonstrate our commitment through our work ethic and sweat. Our passion drives us to give our best and be at our best—*not for our own vanity, but for the benefit of others*. We have high expectations not only for ourselves, but also for others around us. We are never satisfied with mediocrity, and "being average" isn't part of our vocabulary.

> ***Personal Story:*** When I was in middle school, I once got a C on my grade report. My father, a lawyer, had high expectations for his children. Needless to say, he wasn't satisfied with my grade. Knowing his disapproval, I replied, *"Dad, a C means average. A lot of other kids get C's. That's okay, right?"* Without flinching, he looked at me sternly and said, *"Son, you're not average. I know you can do better than that."* That conversation always stuck with me. My dad knew I didn't give my best effort, and he called me out on it. I never received another C on my grade report from then on.

Passion isn't something that can be taught, but it can be "caught," meaning it is infectious and contagious. When we live out our passion for the job, others take notice of our example and start to emulate us. Passionate firefighters are attracted to each other like a magnet to steel.

> **"Remember your first day on the job? Never lose that passion."**

Therefore, let's surround ourselves with passionate firefighters who love the job—let's encourage them and let them encourage us. Let's inspire them and let them inspire us. Let's hold them accountable, and let them hold us accountable.

WHY IS IT IMPORTANT TO BE COMMITTED?

Are we merely *on* the job? Or are we *into* the job?

Firefighters who lack commitment aren't reliable. They cannot be counted on to get the job done. These are the firefighters who hate coming to the firehouse. They lack professionalism, because they complain about the calls they "have" to run, as well as training, cleaning, and public relations events. They are typically the firefighters who love to gossip about other firefighters. They suck the passion and life out of others around them. To them, being a firefighter is nothing more than collecting a paycheck, or if they are a volunteer, it's just about having the title and the t-shirt.

> "If you aren't willing to do the work, don't claim the title."
> -Chief Dennis Reilly

6 STEPS TO STAYING COMMITTED

Don't become an uncommitted firefighter. Adopt the following recommendations on staying committed for the long haul.

1. SET YOUR PRIORITIES.

Whether we are a career or volunteer firefighter, we must realize our time in the fire service will not last forever. Even though the fire service is honorable and we are called to do great things, we must know there is so much more than being a firefighter. That is why we must prioritize what is most important ahead of the fire service.

If you are a person of faith like I am, God must come first. My faith has provided me with a firm foundation and helped me to maintain a perspective of what is truly important in life.

Secondly, our families always take priority over being a firefighter. Without a doubt, the fire service's odd hours, odd shifts, calls, special events, extracurricular meetings, etc., can all pull us away from our families. Being a firefighter means we are part of the fire service family, but we must always put our "first family first."

> "Put your first family first."

When we are home, let's be at home. Let's be intentional about our time with our spouse and children. Let's not let unnecessary distractions come before them. If we take a day off from the firehouse, let's not simply fill it with an extra shift from a part-time job. We must set aside specific time to just be present with our families. Let's establish boundaries that help us balance "the firehouse" and home life.

I have known too many firefighters who have been "married" to the fire department instead of being married to their spouse—meaning they spend most or all of their time taking extra shifts, attending extra training, or doing various fire department events. Some spouses have had to give their firefighters the ultimate ultimatum: *"choose the fire department or choose your family."* According to Anne Gagliano, wife of Captain Mike

Gagliano and co-author of *Challenges of the Firefighter Marriage*, the rate of firefighter divorce is three times that of the general population. We cannot let our marriages become another statistic.

Know the power of *"no."* There is such a thing as being overcommitted. Successful firefighters are full of passion. Our high level of commitment drives us to be all-in, which makes us want to say yes to almost everything: *extra responsibilities, supplemental training and certifications, fire department committees, off-duty public relations events, etc.*

I have had to learn the hard way that saying yes to everything meant I was spreading myself too thin. I became overwhelmed and wasn't following through with everything I said I would do. I had to learn to be able to say no. Once we allow ourselves the permission to do so, we immediately have a newfound sense of freedom. We are able to focus more time and energy on the other things we already committed to doing. The next time someone asks us to participate in something extra and our plate is full, we don't have to reply immediately with a yes or no. Let's tell them that we will give it some thought and get back to them. If our answer is no, we will respond with humility and tact.

> **"It's only by saying no that you can concentrate on the things that are really important."**
> **-Steve Jobs**

2. MAINTAIN A LONG-TERM PERSPECTIVE.

We have all probably heard the old adage: *"Your time in the fire service is a marathon, not a sprint."* When we are new to the job, our passion is overflowing. We are a sponge to learn and do everything. We would love to have the "rookie high" forever. But as time goes on, the natural tendency is for our passion to decrease, or, even worse, we may become disgruntled or bitter.

> **"The pessimist sees difficulty in every opportunity. The optimist sees opportunity in every difficulty."**
> **- Winston Churchill -**

Any firefighter who has been on the job for several years will know that there will be hills and valleys. Excitement for the job will ebb and flow, and this is only natural. There will be negative situations or work relationships that may weigh on us for a while, but a long-term perspective will help us realize that it will eventually get better.

One of the most important things I have learned during my time in the fire service is to never hold a grudge. Two firefighters who may have been the best of friends could become worst enemies who never talk to each other—all over the pettiest negative comment or fault. We must realize that we will work with the same people for 20, 30, or more than 40 years, and when we put up relational barriers we inhibit trust, growth, and teamwork. Let's give everyone the benefit of the doubt, apologize when warranted, and remember: *seek first to understand, then to be understood* (Stephen Covey).

3. KNOW OUR PURPOSE.

Our purpose is the reason we are firefighters. Defining our purpose provides a strong foundation for our entire careers. It will drive and motivate us, helping us to stay committed for our entire careers. No matter the situation we come across, our mission and purpose will guide us through.

> **"The greatest tragedy is not death, but life without purpose."**
> **-Rick Warren**

Ask yourself: *Are you a firefighter for the title, the t-shirt, or the paycheck, or are you a firefighter to make a positive impact and difference in the lives of others?*

We will discuss a firefighter's purpose and mission in greater detail in *Chapter 7 - Conviction*.

4. WORK HARD.

DECIDE + COMMIT + WORK HARD = SUCCEED

There may not be a more simple formula for success. But simple isn't always easy. Choosing to work hard may be a result of our initial commitment. But putting in the work will also have a reciprocal effect on our level of commitment. In other words, *the deeper we commit, the harder we work. The harder we work, the deeper we commit.* The two are intertwined and have a synergistic effect on each other.

Nothing inspires firefighters more than seeing another firefighter work hard. Whether at the fireground or at the firehouse, watching one of our brothers or sisters give their all motivates us to up our game and work harder. Positive peer pressure is always a great thing.

> *Personal Story:* My dad was from a poor family and he admitted to being a very average student growing up. However, he eventually graduated from one of the most prestigious law schools in the country, and then became one of the most successful and respected prosecutors in the city of St. Louis' history. He always told me he wasn't as smart or talented as everyone else. He said he had to work much harder than others to achieve the same amount of success. He taught me everything I know about work ethic, grit, and perseverance—and for that, I am forever grateful.

Lastly, we must remember our work must have purpose behind it. We cannot work ourselves all day and all night if it lacks direction. As Malcolm Gladwell states, *"Hard work is a prison cell if the work has no meaning."* If we are working hard, great. But we must always know our *why*.

> **"Work is what gets the job done. Stop wishing, start doing."**

5. DO NOT WAIT FOR MOTIVATION. BE DISCIPLINED.

While we are discussing motivation, we must be brutally honest: We can't always rely on being motivated. Motivation typically comes from external sources or from our emotions, both of which are unreliable. But we can choose to be disciplined.

Retired U.S. Navy Seal Jocko Willink said it best: *"It's not about motivation. Motivation is a fickle little emotional thing. … It's about the discipline. It's about the discipline of holding the course, knowing what you have to do, and making it happen. Motivation is going to let you down. Discipline will stay by your side."*[3]

> **"Amateurs sit and wait for inspiration—the rest of us just get up and go to work."**
> **- Stephen King -**

Staying committed and being disciplined go hand-in-hand. As disciplined firefighters, we exercise self-control—both over our minds and bodies. Our *champion mindset* doesn't give in to the temptation to be lazy, and we don't take the easy way out. Discipline is like a muscle: The more we exercise it, the larger it grows. The larger it grows, the easier it becomes to put it into action. With discipline we stay the course, even against the odds.

> As disciplined firefighters, we:
> - Wake up early.
> - Show up early.
> - Check the fire truck and its equipment right away.
> - Train, learn, and study every day.
> - Physically train every day.
> - Don't gossip about others.
> - Get back up after failing.
> - *Never give up!*

---

3 "Forget Motivation, You Need Discipline, Navy Seals; Advice." *Motivation Mentalist,* Oct. 31 2016. www,motivationmentalist.com/2016/10/31/forget-motivation-you-need-discipline.

CHAPTER 5

6. CONNECT WITH OTHER COMMITTED FIREFIGHTERS.

We are sadly mistaken if we think we can do it all on our own. Accountability is essential to staying committed. As successful firefighters, we must be willing to surround ourselves with those who exemplify all of the qualities we have discussed in this chapter. We need a support system of committed firefighters who will encourage us on a regular basis. Let's take action and seek out firefighters at our fire departments who we admire. Let's talk to them, ask them questions, train and work out with them.

We may not have a group of such firefighters at our respective fire departments. Don't worry, help is around the corner. Get on social media platforms like Facebook, Instagram, Twitter, LinkedIn, etc., to connect with committed firefighters who love the job. Also, listen to podcasts that have interviews with firefighters who spread a message of honor, passion, and commitment. Learn from their experience and learn about what drives them to be their best.

> "The only way to do great work is to love what you do."
> -Steve Jobs

ACTION STEPS

1. What are your priorities in life? List them below.

2. Write down the names of committed firefighters at your fire department. Make it a point to befriend them and learn from them.

3. Join a committee at your fire department to serve on. If your fire department does not have any committees, start one in a topic you are passionate about *(e.g. fitness/health, training, standard operating guidelines, apparatus, EMS, etc.)*.

4. If you do not listen to podcasts, start listening to one this week. Check out the Firefighter Success Podcast at FirefighterSuccessPodcast.com.

5. With any goal in your life, follow *Firefighter Success'* formula:
DECIDE + COMMIT + WORK HARD = SUCCEED

DECIDE.
COMMIT.
WORK HARD.
SUCCEED.

FirefighterSuccessBook.com

EACH TIME WE FACE OUR FEAR, WE GAIN STRENGTH, COURAGE, AND CONFIDENCE IN THE DOING.

- THEODORE ROOSEVELT -

FirefighterSuccessBook.com

CHAPTER 6
COURAGEOUS

Arland D. Williams Jr. is not a name that necessarily conjures up thoughts of courage and heroism. He may not hold the same historical recognition as such figures of courage like Martin Luther King Jr., Mahatma Gandhi or Nelson Mandela—but as firefighters, we must know his story of courage and sacrifice.

On Jan. 13, 1982, Air Florida Flight 90 departed from Washington National Airport at 4 p.m. With snowy weather conditions and an air temperature of 24 degrees F the plane failed to gain altitude during takeoff. It eventually crashed violently into the 14th Street Bridge, where it struck numerous cars and eventually plunged over the guardrail into the icy Potomac River.[4]

As the plane started to sink into the water, only six survivors were able to get out of the fuselage. Williams and five other passengers waited 20 minutes in the frigid water before a police helicopter was able to come to their rescue. The rescuers dropped a flotation ring and lowered its lifeline first to Bert Hamilton, who was successfully brought to the river's shore. As Williams was clinging to the plane's tail section, he gave the flotation ring to the other survivors. The rescue crew returned and lowered its line directly to Williams, yet he selflessly gave it to the flight attendant, who was brought to safety. The helicopter returned, this time with two ropes, and Williams again gave both

---

[4] "Potomac mystery hero identified". *Toledo Blade*. Ohio. Associated Press. June 7, 1983. p. 1.
https://news.google.com/newspapers?id=MZAxAAAAIBAJ&sjid=uwIEAAAAIBAJ&pg=3020%2C3629030

to the other passengers. The helicopter came back again, and Williams again gave the lifeline to someone else.[5]

As the helicopter was returning to save Williams, the tail section of the plane that he was floating on sank into the water, bringing him with it. Williams drowned in the icy Potomac before he could be rescued. His selfless acts of courage saved the lives of five others that day—*five people who were complete strangers to him.*[6]

Williams' story is a testament to what every successful firefighter must possess—*sacrifice, selflessness, and courage in the face of fear.*

WHAT IS COURAGE?

Courage. Without a doubt, it is one of the quintessential values of successful firefighters. To the public, courage is forever linked with firefighters and what we do. Although movies and television shows may romanticize it, we as firefighters undoubtedly exemplify courage on a daily basis.

In the presence of fear and danger, our courage gives us confidence, poise, and strength. We are able to fully confront our fears, set them aside, and take action. Knowing we are in danger, we have faith that our training and actions will pull us through to a safe outcome. If we were to express courage as a formula, it would be:

CONFIDENCE + FAITH + ACTION = COURAGE

Williams displayed courage by selflessly sacrificing his life for the survival of others. Even though he was in the same life or death situation as the others, he literally gave away all his lifelines before he eventually met his fate. Each time he passed the life ring or rescue line to someone else, he knew his chances of survival decreased. He boldly confronted fear with confidence, faith, and action.

We will discuss three primary types of courage: *physical, character,* and *emotional.* Each is vitally important to our success as firefighters.

PHYSICAL COURAGE

As firefighters, almost all of us are able to easily identify with physical courage. This is what allows us to face danger, the risks associated with it, and then overcome it.

> "Hope is not a strategy. Luck is not a factor. Fear is not an option."
> -James Cameron

---

5 *Ibid.*
6 *Ibid.*

> "You gain strength, courage, and confidence by every experience in which you really stop to look fear in the face. You are able to say to yourself, 'I lived through this horror. I can take the next thing that comes along.'"
> -Eleanor Roosevelt

Physical courage is built on training, knowledge, and experience. Training plays a fundamental part in building our courage when we are confronted with our first working structure fire, our first legitimate vehicle crash, or any other severe call.

Our training gives us mental and physical strength to deal with a situation we may have never experienced before. We feel fear because of the imminent danger that confronts us, or we may not know exactly what to do. But with courage, we find faith and confidence in our training, knowledge, and experience—all of which help us to take decisive action.

Physical courage also comes from trusting in other firefighters who have gone before us. We rely on *their* knowledge, skills, abilities, and experience to carry us through. The Book of Proverbs gives us this wisdom: *"Whoever walks with the wise will become wise, but the companion of fools suffers harm."* We must surround ourselves with solid,

> "Our courage comes from the courage of others."
> -Simon Sinek

successful firefighters who will give us the courage to accomplish great things. There is strength in numbers, and we are capable of so much more with others' knowledge and experience.

CHARACTER COURAGE

We need more than just physical courage to achieve success. Character courage is essential to overcoming the countless moral, ethical, and even mental hurdles that we will face during our careers.

If you're an aspiring firefighter, it is your dream to one day become a professional[7] firefighter. Or maybe you are currently on the job, and it is your goal to one day be promoted to the position of driver, company officer, or even chief officer. It takes courage to dream. It takes even more courage to have the *confidence and faith* to set goals, then *action* and *strength* to achieve these goals.

---

[7] The term professional does not refer to a "career" or "paid" firefighter. Rather, it describes a firefighter who is certified and currently holding a position as a firefighter at a fire department. Volunteer and paid-on-call firefighters are also professional firefighters.

Why? The fear of failure is very real and can be very scary for most firefighters. Failing opens us up to ridicule and embarrassment. This fear can be crippling to firefighters without courage—so terrifying that

> "To climb the ladder of success, you first have to get the ladder off the rack."
> -Chief Jason Hoevelmann

some never take the first step. They are so scared to try because they might become a target for firefighters without character. But as successful firefighters, we have the courage to be open to failure and to be vulnerable—because we know that failure will not hold us back from achieving success.

OVERCOMING FAILURE - A PERSONAL STORY OF COURAGE

Coming up short can be embarrassing. It was for me the first time I took my fire department's promotional exam for the position of lieutenant. Having just six years on the job, I knew that I didn't have the most knowledge and experience (as compared to all of the other candidates). Regardless, I decided that I would study as hard as I possibly could and put my best foot forward.

For the practical evolution at the training tower, my role was to be the first-due company officer, leading my mock crew at a working apartment fire. Before the evolution started, I was to check the truck and crew, ensure their readiness, and communicate my expectations to them. After doing so, I said I was ready to be dispatched for the call. I was nervous, uneasy, and a little nauseous. *But guess what? I mustered up the courage to set these feelings aside and proceed with confidence.*

The scenario started out smoothly. As we arrived on scene, I sized up the structure and gave an accurate initial incident report. As we stepped off the truck, I told my firefighters to pull an attack line and stretch it to the alpha (front) door, bring their tools, and "mask up." The scenario's "occupant" ran up to me as I exited the truck, screaming *"My babies are inside! I don't know where they are! Help them!"* I assured her we would rescue them, and I quickly requested a second alarm response from dispatch for a confirmed "person trapped." (So far, so good.)

During my 360-degree walkaround of the structure, I saw a victim standing at the charlie side second-story window. Instead of telling them: *"Stay right there, we will come to rescue you,"* I radioed my backstep firefighters to grab a ladder and make the rescue. STRIKE ONE. Per my fire department's SOG, if the number of all victims is unknown and there are not enough firefighters on scene to simultaneously perform fire attack and rescue, we must initially execute fire attack.

After my crew entered the building, I directed them to start searching rooms for victims on the way to locating the fire (i.e. its location was still unknown). *STRIKE TWO.* What I should have done was first locate the fire, confine and extinguish it, and then

AFTER THE DUST HAD SETTLED AND MY EMOTIONS HAD SUBSIDED, I HAD TO MAKE A CHOICE. WAS I GOING TO CHOOSE DEFEAT? WAS I GOING TO ACCEPT MY FAILURE AS FINAL? WITHOUT A DOUBT, THAT WOULD HAVE BEEN THE EASY WAY OUT. BUT THE "EASY WAY OUT" IS FOR COWARDS. WE DO NOT TAKE THE EASY WAY OUT. IT TAKES COURAGE TO BRUSH YOURSELF OFF AND GET BACK UP AGAIN. IT TAKES COURAGE TO LEARN FROM FAILURE AND DECIDE TO TRY AGAIN. AND THAT IS WHAT I DID.

FirefighterSuccessBook.com

direct additional arriving crews to perform searches of said rooms. As the old adage goes: *"Put the fire out first and everything gets better."*

Since I initially had my crew rescue the victim and perform searches on the way to locating the fire, I was running short on time. As we made our way through the building, my worst fear came true: *I ran out of time in my scenario and I ended up never locating the fire.* STRIKE THREE.

Needless to say, I felt dejected. I was embarrassed. I was furious with myself. During my study prep, I had gone over scenarios like this dozens of times. I knew my fire department's Structure Fire SOG forwards and backwards. I should have aced this test. But as we can all admit, things don't always go the way we want. Sometimes we think we have all of our bases covered, but then we strike out on a nasty curveball.

> **"Success is not final, failure is not fatal—
> It is the courage to continue that counts."
> - Anonymous -**

I failed, and I knew it. I didn't perform to the level I knew I was capable of, and I was the only person to blame. Needless to say, I did not get promoted during that lieutenant's promotional process.

Courage is what made me take the lieutenant's test again. Courage is what made me learn from my mistakes and study harder than ever before. *Courage is what gave me enough confidence and faith to take action.* By the grace of God and the courage He gave me, I received the highest ranking on my second promotional process, and the rest is history.

5 WAYS TO DEVELOP CHARACTER COURAGE

In Chapter 11, we will discuss character in greater detail. For now, implement the following ways to building character courage.

1. PERSEVERE.

Struggles, hardships, pain, and failure are parts of life, just like joy, winning, and success. It would be great to always win, make a lot of money, and get that promotion without working for it. But this is not how the real world works.

When we fall (and we will), we will brush ourselves off and get back up. When we fail, we will learn from it and be better. We will stand strong in the face of hardship. We will build courage by being resilient during tough times.

> "Our greatest weakness lies in giving up.
> The most certain way to succeed is always
> to try just one more time."
> - Thomas A. Edison -

2. BE HONEST ABOUT WEAKNESSES.

Everyone has weaknesses, and it takes courage to admit it. We cannot be good at everything. If we were, we wouldn't need each other and being part of a crew would be irrelevant. In the fire service, we are part of a team. Whether at the company level, battalion level, shift level, or as a fire department—each of us brings our own strengths and weaknesses. Most importantly, we rely on each others' strengths to accomplish the mission.

Take a second for some honest introspection: What are your weaknesses? *Do you lack confidence? Or perhaps you could be more humble. Do you talk too much? Do you lack discipline? Could you be more compassionate? Maybe your fitness isn't where you want it to be.* Whatever your weaknesses are, be truthful about them. Then have the character courage to improve them. Do one thing every day (big or small) that specifically aims to improve them. Confronting our weaknesses will give us strength.

3. GIVE AND EXPECT ACCOUNTABILITY.

In the same vein as strengths and weaknesses, a firefighter with character courage gives and expects accountability. We have the honesty and courage to hold our team members ac-

> "In the fire service, we are part of a team."

countable to maintaining high standards. Whether it is someone's skill level, knowledge, training, or level of fitness, a firefighter who values accountability will have the tough conversations. We aren't scared to call out our brother or sister when it is necessary—*even if that someone outranks them.*

To become successful firefighters, we must also expect accountability. When we are falling short, we must have the humility (and expectation) to have other firefighters hold us accountable. Let's not forget: One of the most important attributes of successful firefighters is the ability to be coachable. Firefighters who are humble, teachable, and always willing to improve will receive accountability openly.

4. STAND UP FOR WHAT IS RIGHT.

A courageous firefighter will have the strength to stand up for what is right, even if it means standing alone. Unfortunately, the popular choice may at times be the wrong choice. It may feel comfortable and easy to go with the majority, but a firefighter with character courage will have the strength to stand alone, if necessary.

Let's not place too much value in what other people think about us. As long as we are doing the right things for the right reasons, we must brush off their criticism and judgement. We must move forward with confidence, knowing we did the right thing.

Are laziness and mediocrity the standards at our fire station or fire department? *We will stand up.* Are other firefighters singling out or bullying another firefighter? *We will stand up.* Is there corruption within our fire departments, even at the highest levels? *We will stand up.* Remember: We are not firefighters who are on the job to win a popularity contest. We are here to serve others and do what is right.

> "The right choice isn't always the popular choice. The popular choice isn't always the right choice. We will stay true to our values and who we are."

5. CHOOSE WORK.

Laziness is a choice. Mediocrity is a choice. Achieving only the minimum standard is a choice. But these are all poor choices.

Let's have the courage to show up every day, ready to work. Whether it is the firehouse, the fireground, or the gym, it will be our goal to be the hardest workers in the room. There will always be an easy way out. There will always be a path of least resistance. But we know there are no shortcuts to success. It takes courage to get up and work when everyone else may be napping in the firehouse recliners. *We will choose work.*

> "Be the firefighter who everyone has to say: 'Stop working so hard, you're making the rest of us look bad.'"
>
> -Capt. Jim Moss
> **Firefighter Toolbox Podcast 061**

EMOTIONAL COURAGE

VOICES OF EXPERIENCE

Dena Ali
Captain - Raleigh Fire Dept. (NC)
Founder/Director
North Carolina Peer Support

According to the CDC, suicide is a leading cause of death for all Americans, and the third leading cause of death for working age adults.[8] As firefighters, we are not immune to this profound human tragedy. Not only are we dying by suicide but we are also impacted by depression, addiction, sleep disorders, anxiety, and post-traumatic stress. Unfortunately, each of these disorders is associated with an increased risk for suicide, despite the fact that they are common, treatable, and nothing to be ashamed about. This means that when they are addressed, their manifestation along a trajectory towards suicide can be altered.

The stigma associated with these disorders often leads to shame and silence, both of which are associated with increased risk for PTSD and suicide. Shame is a universal feeling that tells us that we are flawed and unworthy.

> "We take pride in being aggressive firefighters on the fireground. We must be equally aggressive when it comes to our mental and emotional health."

Shame coupled with the sense of responsibility that we as firefighters feel towards our communities and crews leads us to fear showing any sort of vulnerability. Because of this, we may hide what we perceive as weakness, and in doing so, exacerbate our suffering while also inhibiting the opportunity for healing. According to University of Texas Psychologist and Air Force Veteran Craig Bryan:

"We train our warriors to use controlled violence and aggression, to suppress strong emotional reactions in the face of adversity, to tolerate physical and emotional pain and to overcome the fear of injury and death, and these qualities are also associated with increased risk for suicide."[9]

Early in our careers, we learn how to remain stoic in the face of danger, but we aren't always taught the importance of sharing our struggles and asking for help when we need it.

---

8 "Ten Leading Causes of Death and Injury by Age Group - 2018." Centers for Disease Control, https://www.cdc.gov/injury/wisqars/LeadingCauses.html.
9 Thompson, Mark. Is the U.S. Army Losing Its War on Suicide?" *Time*, April 13, 2010, http://content.time.com/time/nation/article/0,8599,1981284,00.html.

In his book, *The Beauty of a Darker Soul,* Joshua Mantz explains that trauma is cumulative, and is influenced by everything in our lives that has come before.[10] These include early life experiences, previous trauma, health, relationships, personal factors, and work factors. These are all interrelated and impact how we process the next stressor. Mantz shares that trauma becomes a convenient scapegoat for misunderstood and deeper sources of pain. For firefighters, it is easier to blame stress on the bad calls we run but much more difficult to explore our personal factors and coping skills.

There is a belief that the sole source of stress for firefighters comes from the horrific images and experiences that we accumulate during our careers. For some, this may be true, but for many it's not the only cause. Attributing emotional stress among firefighters strictly to the calls they run can be harmful in that it silences those struggling from other stressors. Those of us dealing with financial problems, relationship issues, or substance abuse disorders may feel that our problems are not "worthy" in comparison to our brothers and sisters who have faced trauma on the job. While each of these sources are different, the detrimental response to the human mind in the form of shame, humiliation, and guilt are all the same.

> "We took an oath to serve and protect others, but what happens when we cannot do that for ourselves?"

We all carry emotional baggage from our experiences, both on and off the job. And we each deal with stress in our own unique ways. In a side-by-side comparison, two firefighters who experience the same exact calls may process them in completely different manners.

The worst thing that we can do is nothing. We cannot bottle everything up inside and be silent. We cannot resort to alcohol or drugs to numb the pain.

A key starting point to healing from trauma and stress is having the courage to give ourselves permission to explore the true source of pain. As Mantz tells us: *"Sometimes this source can be the most microscopic detail of an experience and uncovering those details is often the catalyst that initiates the true healing process."* The next critical step is to realize and admit to ourselves that we cannot cope with these stressors all on our own.

BE STRONG AND COURAGEOUS

It takes courage to be real. It takes courage to be vulnerable. It takes courage to share our emotions and experiences with others.

As successful firefighters, we must have the courage to reach out to others to get help when we need it. We must take a proactive approach by talking to other firefighters

---

10 Mantz, Joshua. *The Beauty of a Darker Soul: Overcoming Trauma Through the Power of Human Connection.* Lioncrest Publishers, 2017, Austin, TX.

about what we are going through, and when needed, we take the next step by connecting with a peer supporter, counselor, therapist, pastor, etc.—*anyone who will listen and provide the help needed.*

Guilt and shame are destructive, but they can be destroyed through the power of human connection and shared experience. Let's trust the power of human connection. Yes, it is scary because we are admitting we may be struggling, and this makes us vulnerable to others' judgement. But let me assure you that reaching out for help is not a sign of weakness. It is a sign of strength and courage.

Lastly, if we observe that one of our fellow firefighters appears to be struggling, let's have the courage to ask them about what they are going through. We may have noticed a change in their behavior, they may become withdrawn, they may send coded messages for help, they could be coming into work under the influence of alcohol or drugs, or they could be going through a divorce, etc.

> We must be very direct with our questions:
> - *"How are you doing?"*
> - *"Are you hurting?"*
> - *"I have noticed you have been [describe the abnormal behaviors you have noticed], and I am worried. Can we talk?"*
> - *"Do you want to kill yourself?"*
> - *"Are you doing drugs?"*
> - *"Have you been drinking?"*
> - *"Do you need help?"*

Don't beat around the bush. Be direct, sincere, and confidential. Let's tell them the changes we have seen and convey our concern with compassion. Let's listen attentively to what they want to share and be empathetic. Being empathetic requires that we eliminate our judgment, our perspectives, and search deep within ourselves for a feeling that can match the one they are experiencing. Our experiences in life with suffering shape who we become as they provide us with opportunities for growth and perspective. There is no new pain in the world, so let's be courageous enough to sit in the darkness with those who are suffering.

If what they are going through is beyond our scope, then we will help get them to more appropriate help. With their permission, let's have the courage to help provide a warm hand-off to more experienced, better-trained care.

EMBRACE THE STRUGGLE

There is no life free of struggle, and while our culture places importance on going at it alone, none of us should have to face our struggles alone. Humans are hardwired for connection. We work in a dangerous profession, but there is nothing more dangerous than suffering alone in silence.

Our experiences give us the ability to connect with others on a much deeper level. This is where we gain our greatest ability to serve others: through the power of human connection and the recognition that we are not alone in our struggles. This is truly the greatest gift we can give another.

Our courage to reach out may save someone else's life, and it may also save our own.

RESOURCES FOR FIREFIGHTER BEHAVIORAL HEALTH

Here is a list of resources that can help firefighters in their time of need.

National Suicide Prevention Lifeline
1-800-273-8255

National Volunteer Fire Council "Share the Load"
NVFC.org/help, Fire/EMS Helpline: 888-731-3473

Rosecrance Florian Substance Abuse Treatment Program for First Responders
Rosecrance.org, Hotline: 866-330-8729

International Association of Fire Fighters Center of Excellence for Behavioral Health Treatment and Recovery
IAFFRecoveryCenter.com, Hotline: 877-776-5494

Safe Call Now
SafeCallNow.org, 24/7 First Responder Line: 206-459-3020

ACTION STEPS

1. Do you have physical, character, and emotional courage? If you are lacking in one of these areas, identify which it is. Take the next steps to growing in this particular area.

2. Think about a specific instance that you failed—either in your personal life or professional life. How did you respond to it? Use the lessons learned from that failure and apply them to the next challenge you encounter.

3. Go to www.nvfc.org/share-the-load-resources/ and print out behavioral health posters and resources for the fire station.

IT IS NOT THE CRITIC WHO COUNTS; NOT THE MAN WHO POINTS OUT HOW THE STRONG MAN STUMBLES, OR WHERE THE DOER OF DEEDS COULD HAVE DONE THEM BETTER. THE CREDIT BELONGS TO THE MAN WHO IS ACTUALLY IN THE ARENA, WHOSE FACE IS MARRED BY DUST AND SWEAT AND BLOOD; WHO STRIVES VALIANTLY; WHO ERRS, WHO COMES SHORT AGAIN AND AGAIN, BECAUSE THERE IS NO EFFORT WITHOUT ERROR AND SHORTCOMING; BUT WHO DOES ACTUALLY STRIVE TO DO THE DEEDS; WHO KNOWS GREAT ENTHUSIASMS, THE GREAT DEVOTIONS; WHO SPENDS HIMSELF IN A WORTHY CAUSE; WHO AT THE BEST KNOWS IN THE END THE TRIUMPH OF HIGH ACHIEVEMENT, AND WHO AT THE WORST, IF HE FAILS, AT LEAST FAILS WHILE DARING GREATLY, SO THAT HIS PLACE SHALL NEVER BE WITH THOSE COLD AND TIMID SOULS WHO NEITHER KNOW VICTORY NOR DEFEAT.

- THEODORE ROOSEVELT -

FirefighterSuccessBook.com

IF YOU LIVE FOR PEOPLE'S ACCEPTANCE, YOU WILL DIE FROM THEIR REJECTION.

- LECRAE -

FirefighterSuccessBook.com

CHAPTER 7
CONVICTION

KNOW THYSELF

What does it mean to have conviction? Better yet, what does it mean to *live* with conviction? Some may say that it comes down to having confidence in yourself. That is partially true, but let's dive in a little further.

Conviction signifies we have a crystal clear vision of who we are, what we stand for, and who we aspire to be.

Conviction represents an unwavering self-belief in everything that we are: *our mindset, attitude, actions, values, knowledge, skills, talents, etc.*

Conviction means we are honest with our strengths, weaknesses, goals, mistakes, etc. Living with conviction shows we are unwilling to compromise when it comes to our core values like integrity, honesty, and trust. It assures a firm foundation and it gives us the confidence to achieve our goals.

In order to know ourselves and what we believe, we must take a moment for some honest introspection.

CHAPTER 7

ANSWER THESE FIVE QUESTIONS:

1. What do I stand for?

2. What do I value?

3. What are my strengths and weaknesses?

4. What mistakes have I made (personally and professionally) and what have I learned from them?

5. What am I unwilling to compromise on?

CREATING A STRATEGIC PLAN

Next, we must create a personal "strategic plan" that identifies and expresses who we are. Yes, strategic plans are used for businesses and organizations, but they are also incredibly useful for people as well. Take some time with this next exercise and give it some deep thought, because it will be worth it. Use the spaces after the questions to write down your responses.

> "When your *why* is big enough, you will find your *how*."
> -Les Brown

1. **Mission:** What is my purpose?

2. **Vision:** Who do I aspire to be? What do I aspire to do with my life?

3. **Core Values:** What are my core values? (*Use the Core Values List on page 86 to help identify yours. Circle and list at least five that are most important to you.*)

4. **Goal Setting:** What are my short-term and long-term goals? (*Use the SMART method of goal-setting: Specific, Measureable, Attainable, Relevant, Timebound.*)

5. **How will I achieve my goals?** What are very specific things I will do in the short-term (3–6 months) and in the long-term (1–5 years) to make tangible progress?

CHAPTER 7

CORE VALUES

HONESTY	POSITIVITY	PROBLEM SOLVING
RESPECT	PERSEVERANCE	TEAMWORK
TRUST	COURAGE	PROFESSIONALISM
INTEGRITY	COMPASSION	FAIRNESS
CHARACTER	QUALITY	GRATITUDE
LOYALTY	CONFIDENCE	HOPE
SERVICE	HUMILITY	INNOVATION
FAITH	PASSION	INCLUSIVE
ACCOUNTABILITY	CREATIVITY	LOVE
GENEROSITY	TOUGHNESS	PATIENCE
DEPENDABILITY	DISCIPLINE	EXCELLENCE
RELIABILITY	PERSISTENCE	EMPOWERMENT
CONSISTENCY	HARD WORK	CURIOSITY

PERSONAL STRATEGIC PLAN EXAMPLE

1. **Mission:** To serve others to the best of my abilities, both in my personal life and as a firefighter.

2. **Vision:** To achieve my full potential and success as a spouse, parent, and firefighter.

3. **Core Values:** Integrity, honesty, humility, ownership, and service.

4. **Goal Setting:**
   a. **Short-term Goal:** To obtain the Fire Instructor I and Fire Officer I certifications in the next six months.
   b. **Long-term Personal Goal:** To own a home and have a thriving marriage with three children by the time I am 35 years old.
   c. **Long-term Professional Goal:** To obtain the rank of company officer by the time I have 10 years on the job.

5. **How will I achieve my goals?**
   a. I will enroll in Fire Instructor I and Fire Officer I classes as soon as I can.
   b. I will save $200 each paycheck to put towards a down payment of my future home.
   c. I will find a company officer to be my mentor, one who will help me specifically prepare for the position.

Once we have created our personal strategic plan, we can move forward with conviction, because we know our purpose, who we are, our core values, and our goals. As months and years go by, it is perfectly acceptable to adjust and update our strategic plans.

BELIEVE IN YOURSELF WHEN NO ONE ELSE WILL

If we do not know who we are and what we stand for, others will decide for us. If we are unsure of what we believe, we will be unsure of what to say and do. We must have an unwavering belief in ourselves to achieve success.

There will be negative people and distractors at every step in our journey, criticizing every single thing we say and do. We cannot give in to them, and we cannot let them walk all over us. Remember what we discussed in *Chapter 3 - Champion Mindset:* We will stay positive and we won't let anyone tear us down.

> "You have enemies? Good.
> That means you've stood up for
> something, sometime in your life."
> - Winston Churchill -

5 WAYS TO DEAL WITH NEGATIVE PEOPLE

1. **Know that it's not us, it's them.** Negative people tend to project their insecurities on those who make them feel insecure (i.e. successful people). If what they are saying is completely false, disregard what they say. If there is truth to their criticism about us, let's take it and use it to improve.

2. **Brush them off.** Negative people criticize others just because they want to elicit a reaction 99% of the time. Let's maintain our poise, be confident, be respectful, and do not let them get under our skin.

3. **Do not stoop to their level.** When others criticize us without reason, we may feel the urge to give it right back to them. As difficult as it might be, let's take the high road.

4. **Use their criticism as fuel.** David Brinkley offers this priceless advice: *"A successful man is one who can lay a firm foundation with the bricks others have thrown at him."* The critics will throw bricks at us. We can either let them weigh us down, or we can use them to stand upon to rise a little higher.

5. **Surround ourselves with the best people.** As a counterpoint to the negativity, let's surround ourselves with positive, encouraging, successful firefighters. Let's emulate their behavior and actions. Let them build us up in a way that will help us succeed.

> "A lion doesn't concern itself
> with the opinion of sheep."
> - George R.R. Martin -

TAKE ACTION

Conviction means nothing without action.

We may say our core values are integrity, honesty, trust, etc., but if we do not live them out on a daily basis our beliefs are irrelevant. Whether we like it or not, our actions always reflect our thoughts, beliefs, and attitudes. If we do the opposite of what we say we believe, then we are hypocrites. We will lose all credibility and respect if we do not practice what we preach—especially those of us who are fire department leaders.

We will be firefighters with conviction and courage, living out our beliefs through action, intentionality, and purpose.

ACTION STEPS

1. Answer the questions at the beginning of this chapter.

2. Create your personal strategic plan.

3. Identify the negative people in your life. Minimize your contact with them. If you cannot do so, come up with specific strategies to deal with them (e.g. setting boundaries, "killing them with kindness," etc.)

4. Identify two or more successful firefighters in your fire department with strong conviction. Build relationships with these firefighters. Encourage each other and hold each other accountable.

KNOW THE JOB.
DO THE JOB.
LOVE THE JOB.

FirefighterSuccessBook.com

CHAPTER 8
COMPETENT

TRAINING IS THE FOUNDATION

A competent firefighter *knows the job.* As competent firefighters, we have complete confidence in our knowledge and skills, and our citizens have complete confidence in us.

We are successful throughout our careers because we know training is the foundation—not only to our success but also to our crew's success and to the mission's success. We take a proactive approach to everything we do, because we know being prepared is vital to successful outcomes on every call. We take our fitness seriously, because being physically prepared is a requirement of every firefighter.

> "The fireground isn't our dress rehearsal. We train to win."

> "Our titles, t-shirts, and tattoos don't make us firefighters. Our training does."

We train daily to ensure competency and operational readiness. This is what our public expects and this is what they deserve. Our aim is to master the craft—not for our vanity or to know more than everyone else. Rather, we train hard because we want to be the best that we can possibly be.

This self-investment will pay huge dividends for our crew, our fire department, and our citizens.

> "Are you a firefighter? Do you want to be a firefighter? Then train like a firefighter is supposed to train."
> - Dan Kerrigan & Jim Moss -
> *Firefighter Functional Fitness*

Capt. Robby Owens of the Henrico County Division of Fire (Virginia) offers a great mindset for daily training. Every time he and his crew are on duty, it is their goal to achieve three hours of training. He breaks this down into three specific areas:

- One hour for hands-on training
- One hour for personal reading
- One hour for physical fitness

When we adopt and apply this methodology, we are developing our skills, our minds, and our bodies. Adopting this process will make us harder to kill on all fronts.

THE 3 R'S OF TRAINING FOR COMPETENCE

For firefighter training to be effective and long-lasting, it must follow the 3 R's — *relevance, realism, and repetition*. Let's examine how we can apply each of these principles to our training regimens.

1. RELEVANCE

Our training must be relevant to what we actually do on the fireground. If our training does not directly apply to what we do, then what is its purpose? Too many times, firefighters want to "what if" everything to death, and, therefore, come up with crazy, off-the-wall scenarios that have a 0.01% chance of actually happening. Instead of focusing on a "cool new technique" and adding "another tool for the toolbox," let's first focus on the basics: *stretching hose, flowing water, ground ladders, ropes and knots, hand tools, forcible entry, search and rescue, etc.*

> "Master the basics until they become advanced."
> -Kevin Shea
> Firefighter (Ret.) - FDNY

The basics are always best. Some firefighters may write this statement off as cliché, but the basics never get old. The basics are what build a solid foundation for mastery. There are so many subdisciplines in the fire service that the concept of mastering the basics will undoubtedly take up enough of our time.

2. REALISM

Too many times, I have witnessed firefighters who train in completely unrealistic environments, with circumstances that will not sufficiently prepare them for fireground success. Typically, firefighters will choose to wear the least amount of personal protective equipment during training to avoid being uncomfortable. Whether they are trying to avoid getting hot, sweaty, or general discomfort, they are only doing a disservice to themselves when "game day" comes.

> **"I regret my training," said no firefighter ever.**

Should every hands-on training evolution be performed in full PPE while being "masked up?" No, of course not. For example, it would be impractical and unnecessary for a firefighter to be in full PPE for a through-the-lock forcible entry class. However, training evolutions that are physically demanding (e.g. ground ladders, hose deployment, conventional forcible entry, search and rescue, etc.) must be practiced with varying levels of PPE. It is acceptable to start the initial "reps and sets" with just bunker pants, gloves and a helmet. However, the end goal for optimal proficiency, confidence, and competence is to perform them exactly how we do on the fireground—in full gear, with an SCBA, masked up, and "on-air."

We have all heard the fire service motto: *"Train like you fight. Fight like you train."* In other words, how we practice is how we will perform. If we always train in the minimal amount of our PPE to avoid discomfort, executing in full PPE on the fireground will be much more difficult. Most firefighters would be shocked to know that wearing full PPE and breathing from our SCBA reduces our cardiovascular capacity (output) by 22%, and also reduces our muscular power output by 20%.[11] Furthermore, a firefighter wearing their full PPE reduces their maximum physical performance by 25%, while concurrently increasing metabolic expenditure by 50%.[12]

The next time we perform rigorous hands-on training, let's consider these sequential levels of using our firefighter PPE:

> 1. Duty uniform
> 2. Turnout coat and pants, helmet, and gloves
> 3. Add an SCBA
> 4. Mask up and go "on-air." (i.e. "full PPE")

---

[11] Eves, Neil, et al. "The Influence of the SCBA on Ventilator Function and Maximal Exercise." *Canadian Journal of Applied Physiology*, November 2005.
[12] Kerrigan, Dan and Moss, Jim. *Firefighter Functional Fitness*. Firefighter Toolbox LLC, 2016, p. 148.

OUR DEPARTMENT TAKES 1,120 CALLS EVERY DAY. DO YOU KNOW HOW MANY OF THE CALLS THE PUBLIC EXPECTS PERFECTION ON? 1,120. NOBODY CALLS THE FIRE DEPARTMENT AND SAYS, SEND ME TWO DUMB-ASS FIREFIGHTERS IN A PICKUP TRUCK. IN THREE MINUTES THEY WANT FIVE BRAIN-SURGEON DECATHLON CHAMPIONS TO COME AND SOLVE ALL THEIR PROBLEMS.

- CHIEF JOHN EVERSOLE (RET.) -
CHICAGO FIRE DEPARTMENT (PARAPHRASED)

FirefighterSuccessBook.com

A WORD OF WARNING ABOUT COMPUTER-BASED TRAINING

It is undeniable that we live in the age of technology. Computer-based training is becoming more popular, and firefighters are using their smartphones and tablets more than ever. Without a doubt, this medium is convenient and efficient to deliver training to a large group of firefighters. It is not, however, always effective. Moreover, I have witnessed firefighters take an online class/module together, only to cheat by sharing answers with each other. Such training lacks accountability and does not accomplish what it is intended to do.

For the firefighter who needs to take basic hazardous materials continuing education hours, computer-based training can be acceptable. However, there will never be a substitute for the fundamentals of firefighting—*stretching lines, throwing ladders, and forcing doors.*

> "Firefighting is a physical job, and no amount of training videos will ever take the place of gearing up and getting sweaty."

If we find and share a quick "how-to" video on social media for a better way to deploy hose or carry ladders, that is great. But let's make sure that we actually do some hands-on training after we watch it to verify that it will work for us. With social media, it can be the "Wild West" when it comes to firefighter training—There are some legitimate training techniques with proper execution, and then there are some that should never have been posted for the world to see. Again, if we see something we think we can use to improve our firefighting skills, we must vet it ourselves on the training ground.

3. REPETITION

In his book, *Outliers,* Malcolm Gladwell proposed that *"10,000 hours is the magic number of greatness."* Using Bill Gates and The Beatles as two of his examples, Gladwell stated that those who want to "master their craft" need to put in 10,000 hours of practice. Whether we believe the 10,000-hour rule or not, his premise is absolutely true—to achieve competence and mastery, we must invest a huge amount of our time and effort. *We must train, train, train ... And then train some more.*

> "Sets and reps—that's what our public expects."

While we are mastering our position, we must do two things. First, teach everything we know to firefighters with less knowledge and experience. Then learn the position that is immediately above our current one. With regard to the former, when we train and mentor others, we become better at our current position. As the old adage goes, *"Teaching is learning twice."* With regards to the latter, we never know when we will be called to fill in for someone who is at the next level. Even if being promoted is not one of our goals, we must be prepared for unforeseen future opportunities.

> "Master your position. Train the rookie.
> Know the position above your own."

101 IDEAS FOR HANDS-ON TRAINING

The following 101 training ideas do not make up a comprehensive list of every type of firefighter hands-on training. Use this list to spur more ideas for your training and your crew's training. Adapt them to your fire department's procedures and practices.

1. Attack hose loads, deployment, and reloading
2. Attack hose stretching (e.g. up and down stairs, around corners and obstacles)
3. "Courtyard Stretch" for long stretches (e.g. 2.5" hose reduced to 1.75" attack hose)
4. Charged attack hose advancement (e.g. single and multiple firefighters, walking, crawling, "clamp and slide," up and down stairs, etc.)
5. Flowing water from attack hose (e.g. different diameter hoses, single and multiple firefighters, different positions: sitting, kneeling, standing, flowing while moving, etc.)
6. Timed attack hose evolutions: How long does it take your crew to stretch an attack line and flow water after the driver sets the parking brake? (i.e. "2-minute drill," NFPA 1410 drills, etc.)
7. Master stream operations (e.g. deck gun, Blitzfire, aerial ladder pipe)
8. Transitional attack
9. Attic fire attack (i.e. eave and gable vent attack methods)
10. Standpipe operations
11. Flowing water with foam (e.g. proper setup, application, calculations, etc.)
12. Compressed Air Foam Systems operations
13. Alternative nozzles (e.g. piercing nozzle, attic spikes, Bresnan/cellar nozzle, Fognail/AttackSpike™, etc.)
14. Vehicle fire extinguishment procedures (e.g. semi-trucks, hybrid, electric, C.N.G.)
15. Hydrant connections and operations
16. Supply hose lays (e.g. forward, reverse, split, "no-water"/hydrant operations, "rural hitch", etc.)
17. Pump operations: Pumping from the firetruck's tank
18. Pump operations: Pumping off of a hydrant
19. Pump operations: Supplying water to another firetruck
20. Pump operations: Supplying water to a fire department connection (i.e. standpipe operations, sprinkler system, etc.)

21. Pump operations: Drafting
22. Aerial ladder capabilities (e.g. length, load capacity)
23. Aerial ladder operation by driver/operator (e.g. extension, raising, retraction, bedding it, scrub angles, short-jacking the outriggers, etc.)
24. Aerial ladder: Ascending in full PPE, carrying tools, carrying a roof ladder, etc.
25. Aerial ladder: Accessing commercial roofs with tall parapets (e.g. using the roof ladder connected to the tip of the aerial ladder)
26. Aerial ladder: High-angle victim removal with a Stokes basket and rope systems
27. Aerial ladder: Use of ladder belts/carabiner to clip in
28. Aerial ladder: Firefighter operating the controls from the aerial ladder's tip
29. Aerial ladder: Driver/operator and firefighter operating the aerial's nozzle both from the turntable and the tip of the ladder
30. Ground ladders: Reviewing the types, lengths, locations, capabilities on your truck
31. Ground ladders: Lifts (2 firefighters, 1 firefighter, straight-arm, low-shoulder, high-shoulder)
32. Ground ladders: Carries (same as lifts)
33. Ground ladders: Raises (single and multiple firefighters, hands on rungs, hands on beams, side beam raises)
34. Ground ladders: Drags (also with tools and saws)
35. Ground ladders: Footing on uneven ground (e.g. using wedges and cribbing, a halligan on a slope, digging out the dirt)
36. Ground ladders: Accessing windows (2nd, 3rd, 4th floors)
37. Ground ladders: Accessing balconies and roofs
38. Ground ladders: Horizontal ventilation of windows
39. Ground ladders: Operating on ladder with leg lock and/or knee lock
40. Ground ladders: Window bail out (i.e. elbow lock and rotate)
41. Roof ladders: Placement
42. Roof ladders: Vertical ventilation with roof saw
43. Roof ladders: Vertical ventilation with hand tools
44. Search and rescue: Basic residential search
45. Search and rescue: Large commercial search
46. Search and rescue: Vent-Enter-Isolate-Search (VEIS)
47. Forcible entry: Inward-swinging door, outward swinging door, single and multiple firefighters
48. Forcible entry: Commercial through-the-lock techniques
49. Forcible entry: Defeating chains and padlocks

50. Forcible entry: Using rotary saw for cutting door/window bars, locks, chains, carriage bolts and drop bars
51. Forcible entry: Using rotary saw to cut garage doors and commercial roll-down doors
52. Firefighter personal protective equipment: Donning full PPE in less than 60 seconds
53. Self-contained breathing apparatus: Basic operations (e.g. donning, heads-up display, PASS alarms, etc.)
54. Self-contained breathing apparatus: Masking up quickly (i.e. 15 seconds or less while keeping gloves on)
55. Self-contained breathing apparatus: Refilling a cylinder from a cascade system
56. Self-contained breathing apparatus: Rescue breathing with other firefighters
57. Self-contained breathing apparatus: Rapid intervention pack familiarization and proper use of all connections
58. Self-contained breathing apparatus: Donning with seatbelt on while apparatus is in motion
59. Self-contained breathing apparatus: Emergency air consumption techniques (i.e. "low-air" drills, also incorporate "calling the mayday")
60. Self-contained breathing apparatus: "Firefighter Dodgeball" (dodgeball while in full PPE and "on-air" for the purposes of building PPE confidence, air consumption, and physical fitness)
61. Self-rescue: Forcible egress (i.e. wall breaches, confined space maneuvers, maneuvering through the rungs of a roof ladder)
62. Self-rescue: Using the hoseline to find the exit (i.e. "smooth, bump, bump to the pump")
63. Self-rescue: Wire disentanglement
64. Self-rescue: Using personal bailout systems
65. Self-rescue: Elbow/leg lock and hang from a window (i.e. when no ground ladder is present at the window)
66. Rapid intervention crew operations: Gathering the correct tools, size-up, forcible entry, ground ladders, radio communications procedures
67. Rapid intervention crew operations: Calling a mayday (i.e. "Who, what, where," activating the PASS alarm, etc.)
68. Rapid intervention crew operations: Rescuing an incapacitated firefighter (i.e. firefighter lifts, drags, and carries)
69. Rapid intervention crew operations: The Denver Drill
70. Rapid intervention crew operations: The Pittsburgh Drill
71. Rope rescue: Basic knots (e.g. half hitch, clove hitch, figure 8 family, safety, butterfly, bowline, water knot, etc)

72. Rope rescue: Lowering and hauling systems (e.g. setting up a belay, using the MPD, setting up a 3:1 haul and a 5:1 haul, change of direction, prusiks and Gibbs ascenders, converting from a lowering system to a hauling system, etc.)
73. Truck company responsibilities: Ladders, Overhaul, Ventilation, Entry (forcible), Rescue, Search, Utilities (LOVERS-U)
74. Auto extrication: Rescue tool techniques and operations, cutting, dashboard rolls, etc.
75. Auto extrication: Lifting (air) bags (i.e. hand-made bowling ball maze with 4x8 wood prop)
76. Auto extrication: "Rescue Tool Jenga" with 4x4 wood cribbing and the spreader tool
77. Auto extrication: Stabilization with Paratech™ and/or Rescue 42™ struts
78. Auto extrication: Cribbing
79. Auto extrication: Hybrid, electric, and compressed natural gas vehicles
80. Motor vehicle accidents: Apparatus positioning for highway incidents
81. Firefighter confidence course: 5-10 basic firefighter skills done in sequence (e.g. "Work performance assessment" in full PPE)
82. Thermal imaging camera: Capabilities, limitations, size-up, color palette, search and rescue
83. Fire behavior: Flashover, rollover, backdraft, smoke explosion, flow path (neutral plane, bidirectional, unidirectional), thermal ballast, etc.
84. Elevator rescue operations
85. Trench rescue operations
86. Confined space rescue operations
87. Building walk-throughs, inspections, and general building construction (i.e. Type I-V)
88. Drivers training: Driving, backing, parking, positioning the truck for specific scenarios
89. Area familiarization: Streets, high-hazard occupancies, etc.
90. Fireground rehabilitation procedures
91. Cancer prevention procedures (e.g. fireground decontamination)
92. Portable radio operations: Switching channels, scanning, emergency procedures/communications
93. Hazardous materials response and gross decon setup
94. Radio reports (e.g. size-up and "initial incident reports," "conditions-actions-needs" reports)
95. Multicompany structure fire scenarios
96. Hand tools: Inspection, maintenance, terminology, history, tuning, and modifications
97. Fire extinguishers: Using and refilling (e.g. water extinguisher)
98. Scavenger hunt for equipment on the fire truck
99. Incident command/NIMS/Blue Card Command™

100. Post-incident analysis: Reviewing important and unique calls as a group after the call is over
101. Physical fitness

A QUICK WORD ON MASKING UP

We can tell a lot about a firefighter's level of training by how long it takes for them to "mask up." There are some who fumble with their equipment, cannot find a glove, perform the sequence incorrectly, etc. Every inconsistency adds to their total time, frustration level, and overall stress.

One of the drills I like to do is masking up and going "on-air" for speed. I encourage my firefighters to mask up while keeping their gloves and helmet on. Those who have trained on this know that it can save a great deal of time, as compared to a firefighter who has to take their helmet off, set it on the ground, take their gloves off, don their facepiece, connect the SCBA regulator, pick their helmet back up and put it on, click their chinstrap, put their gloves back on, etc.

> **"Proper Planning and Preparation Prevents Piss Poor Performance."**
> -The 7 P's of the British Army

The following is a quick drill and sequence for masking up while keeping our gloves on.

- Don your full PPE ensemble: pants, boots, coat, SCBA, gloves, helmet, radio, flashlight, etc.
- Ensure your SCBA face mask is connected to the regulator. The top straps of the face mask should be at their "final tightness." The bottom straps should be all the way extended.
- Ensure your helmet's chin strap is buckled.
- Gloves are on both hands.
- **Gloves-on Mask-up Sequence:**
    a. Turn on your SCBA cylinder all the way on with your right hand.
    b. While you are doing this, loosen your helmet's chin strap with your left hand.
    c. Remove your helmet from the top of your head with your left hand and rest it behind your head or to your side. The chin strap will be resting against the front of your neck. The helmet does not get separated from your body. It does not get set on the ground.

d. Now grab your face mask and don it with both hands grabbing the lower straps and their buckles. You will hook the face mask's netting on the back of your head, and slide the actual mask over your face. Tighten down the lower straps to achieve a tight seal.

e. Then slide your hands back along your neck to "hook" your hood with both thumbs. Pull it over your scalp and over the rubber pieces of your mask, covering all exposed skin.

f. Lastly, place your helmet on top of your head and tighten your chin strap.

Once we become proficient in this method, it can be done as quickly as 15 seconds or less. Turning our SCBA cylinder on as we exit the fire truck at the fireground will save us another few seconds of time. Disclaimer: The first several times a firefighter attempts this method can be awkward and even frustrating. But with practice and consistency, our efficiency and time will improve.

KITCHEN TABLE TRAINING

Not all training happens on the drill ground. Due to inclement weather, a very busy schedule, or other extenuating circumstances, we may choose to do indoor training. On my crew, we do "Kitchen Table Trainings" almost every time we work. More often than not, we complete them in the morning while our crew eats breakfast.

Our Kitchen Table Trainings consist of five questions that relate to different disciplines within our profession. Typically, we have questions about firefighting, policies/SOGs, area familiarization, "miscellaneous," and EMS (since we are also paramedics). I give the crew 5—10 minutes to write down their answers, then we discuss the correct answers after everyone is done.

KITCHEN TABLE TRAINING EXAMPLE:

1. **Fire:** *How many portable ladders do we have on our truck? What lengths are they?*
2. **Policy/SOG:** *If you need to call in sick, how soon before your upcoming shift are you required to do so? Who do you notify?*
3. **Miscellaneous:** *You are first to arrive on the scene of a working structure fire. What are the items that should be included in your radio report?*
4. **Area familiarization:** *How many elderly care facilities are present in our district? Which have elevators?*
5. **EMS:** *What are the indications and dosages for Atropine?*

CHAPTER 8

As one can see, the questions are not overly complicated. However, they do produce opportunities for great discussion. They also can identify areas of weakness that may need to be addressed through additional training. Since these are so simple, anyone can create them—it does not have to be the company officer or training officer. Truth be told, having a newer member create and deliver them provides a distinct perspective and training opportunity for the entire crew. Consider implementing them with your crew as soon as possible.

KITCHEN TABLE TRAINING - 002

FIRE - What are the three working ends of the halligan bar called? What is a halligan's standard length and why?

EMS - What are the two approved sites for intraosseous needle insertion?

POLICY/P.O.M./C.B.A. - What are the District's official "work hours?" (e.g. For training, truck checks, etc.). Weekdays? Weekends? Holidays?

MISCELLANEOUS - If you are attempting transmit radio traffic over your portable radio, but cannot due to a poor signal on your TAC channel, what should you do?

AREA FAMILIARIZATION - If you are driving southeast on Kehrs Mill Rd. From Clayton Rd., give two cut-through options to Manchester Rd. If you need to get to New Ballwin Rd.

JPM 5/2018 - 002

Kitchen Table Training Example

THE 3 PILLARS OF BUILDING COMPETENCY

As competent firefighters, we understand our initial "firefighter certification" is the minimum standard to our body of training. Graduating from the fire academy is only the beginning of a lifelong process to continuous improvement. In addition to basic training, successful firefighters also build competency through education and experience—the two remaining pillars. Both are vital to building a well-rounded perspective and knowledge base.

> "Eyes, ears, and experience will equal your education on this job!"
> -Lieutenant Mike Ciampo, FDNY

Source: Kobziar, Leda N., et. al. "Challenges to Educating the Next Generation of Wildland Fire Professionals in the United States." Journal of Forestry -Washington-, vol. 107, no. 7, Oct. 2009, pp. 339-345

We have discussed the importance of frequent training to achieve strong firefighter skill sets. But training must be supported with education and experience. Chief Jim Broman of the Lacey Fire District (Washington) shares an interesting comparison on *training* and *education*. He offers that training is *"anchored in the past, what to do, centered around job skills, application, and confronting the known."* On the other hand, education is *"what to be, focused on the future, built around life skills, and confronting the unknown."*

> "Training produces the skills. Education provides perspective. Experience instills true knowledge."
> -Anonymous

Education can be both formal and informal. Formal education typically takes place in a "brick and mortar" classroom or via online classes. Formal education helps build a greater perspective, since the firefighter is taught multiple viewpoints, including lessons learned from fire service history.

For an undergraduate degree, a college fire science program is incredibly beneficial because it takes the student through multiple fire service subdisciplines: fire behavior, building construction, building codes, fire service history, fire investigation, officer development, and much more. It also builds discipline and organizational skills, since the student must "show up," do the course work, study, meet deadlines, and prove their learning through examinations. Additionally, the general education courses (e.g. math, science, composition, etc.) make us more well-rounded, give us a broader perspective, and help develop critical thinking skills. However, we must state that obtaining a college degree does not automatically make us competent, but it is a key element of the process.

In addition to a college degree, consider these other educational opportunities for firefighters:

- Conferences - local, regional, and national
- National Fire Academy
- Online webinars and classes (beyond what is required by our fire departments)

Informal education is equally important as its counterpart. This type of education typically happens during a person-to-person transfer, whether it is between a mentor and mentee, company officer and firefighter, or even peer to peer. Informal education can happen day or night, in the firehouse or at the tailboard after a fire, on or off duty.

When we are new to the job, we are like a sponge: always hungry for more knowledge. If we have a good officer and crew, they will educate and guide us on what to do and what not to do. They will teach us the unwritten rules and culture of the fire department. They will show us how to work smarter, not harder. Pay attention, because this type of education is priceless.

> "Learn one thing every one shift for your whole career, and one time, at any one call, you may make a difference in one life. That one life may be your own!"
> - Capt. John Shepherd -

THE ROLE OF EXPERIENCE

The third pillar of building competency is *experience*. We can train and we can study, but until we put it all into practice, it is just theory. We may attend a class or watch a webinar on fire dynamics, but until we actually experience the heat and observe a smoke explosion firsthand, we are only partially educated.

There is a difference between throwing a ground ladder on a flat drill ground when it is 70 degrees and sunny—compared to doing it on a bitterly cold and rainy night at a working house fire with our adrenaline in overdrive and our heart beating through our chest. Of course, the hours spent in training will develop a strong base of skills, but only the fireground can affirm those skills through real-world experiences.

> "Experience is the best teacher: It gives you the test first and the lesson later."
> -Oscar Wilde
> *Paraphrased*

Whether we are the rookie, veteran firefighter, company officer, or chief, we all learn through experience. Either in the firehouse or at the fireground, both "good" and "bad" experiences teach us what books and classrooms sometimes cannot.

> "The only bad experience is that which we refuse to learn from."

As a company officer, I learn every day. I have taken numerous leadership classes and read countless books on the topic—all of which have been beneficial. But leadership solely learned through these mediums will not suffice. As firefighters and fire officers, we must "get our hands dirty," using experience as our primary teacher. A fire service leader must be tested by fire—sometimes in the literal sense of the word, and sometimes figuratively. I have learned my most important leadership lessons through having to make tough, and sometimes unpopular decisions. Sometimes I made the best decision possible, and sometimes I should have chosen a better option. Either way, I have learned to be a more competent leader through all of my experiences.

LEARN FROM THE EXPERIENCE OF OTHERS

What is better than learning from our own experience? Learning from others' experiences. As Gina Greenlee tells us: *"Experience is a master teacher, even when it's not our own."* We discussed the importance of informal education and how more experienced firefighters must teach us "the way" when we are new. Most of what they pass on comes from their experiences on the job. Such knowledge should be treasured up as gold—It is invaluable because it helps us avoid making unnecessary mistakes. It will help us to operate more effectively, efficiently, and safely.

Consider these 5 foolproof ways to learn from others' experience:
1. Have a mentor.
2. Ask experienced firefighters and officers plenty of questions:
   What would you do if...? What would you have done? Have you ever come across this situation?
3. Observe others' actions, behaviors, successes, mistakes, etc. The benefits of simple observation are quite underrated.
4. Read books, especially biographies.
5. Listen to podcasts, video interviews, TED Talks™, etc.

"THERE ARE NO MISTAKES IN LIFE, ONLY LESSONS. THERE IS NO SUCH THING AS A NEGATIVE EXPERIENCE, ONLY OPPORTUNITIES TO GROW, LEARN AND ADVANCE ALONG THE ROAD OF SELF-MASTERY. FROM STRUGGLE COMES STRENGTH. EVEN PAIN CAN BE A WONDERFUL TEACHER."

— ROBIN SHARMA —

FirefighterSuccessBook.com

TRAIN HARD, TRAIN TRUE, TRAIN OFTEN

In the words of Fire Chief Forest Reeder: *"Great training has to do one of four things: Make us better, faster, safer, smarter."* We cannot take our training for granted. We must seize every opportunity to become better. Whether it is on the drill ground, classroom, or on the fireground—our training, education, and experience are what will develop us into competent, successful firefighters.

Whether we have been to 1,000 fires or just one is irrelevant. What matters most is that we are fully prepared for the next one. As Captain Tom Brennan (FDNY) has shared, *"You can never learn too much about a job that can kill you."*

Let's train hard, train true, and train often.

ACTION STEPS

1. From the "101 Ideas for Hands-on Training" list above, choose three topics you will train on over the next month. Invite your crew to train with you. Keep the momentum going by choosing three more topics for the following month.

2. Train on the "Mask-up Drill" described above. How fast can you mask up?

3. Create and deliver a Kitchen Table Training for your crew.

4. Read one book per month for the purpose of self-improvement. Don't limit your choices to only firefighter training. Other topics include leadership, mindset, physical fitness, etc.

5. If your health and fitness are lacking, start taking better care of yourself. Eat better, exercise, and get more sleep. Go to FirefighterFunctionalFitness.com for a simple and practical approach to firefighter fitness.

6. Take at least one specialized firefighter certification or training per year. Here is a list to spur on ideas:

- Driver/operator
- Fire instructor
- Fire officer
- Hazardous materials technician
- Rope rescue technician
- Confined space technician
- Fire investigator
- Fire inspector
- Vehicle extrication
- Wildland firefighting
- Aircraft rescue and firefighting (ARFF)
- Dive rescue specialist
- Swiftwater technician
- Boat operator
- Ice rescue
- And more …

IN ALL THINGS, THE DETAILS MATTER.

FirefighterSuccessBook.com

CHAPTER 9
COMPREHENSIVE

One essential quality of successful firefighters that has stood out to me over the years is that they are comprehensive in everything they do. Not only are they *thorough* in their approach to the job, but they are also *well-rounded* in their preparation and the execution of their duties.

BE A WELL-ROUNDED FIREFIGHTER

Let's be honest, the days of being "just a firefighter" are long gone. As modern-day firefighters, we serve our fire department and the public in multiple disciplines. We must be *jacks of all trades*, meaning that we must be well-rounded and comprehensive in all areas of the job.

> **"We don't just fight fires anymore."**

Consider all of the various roles and specializations that we may be called to perform:

- Structural firefighter
- Wildland firefighter
- Aircraft rescue and firefighting
- Vehicle extrication
- Hazardous materials technician

109

- Rope rescue technician
- Confined space technician
- Swiftwater technician
- Dive rescue specialist
- Boat operator
- Ice rescue
- Bike/trail response team
- Fire code building inspection
- Fire investigation
- Community liaison/public relations and education
- Emergency medical technician
- Tactical medic
- And more ...

Truth be told, firefighters don't just fight fires anymore. Being part of an all-hazards fire department requires us to be comprehensive in our approach to emergency response. So, the million dollar question must be asked: *"How can we possibly do all of these things?"*

I will be honest, and there will be some that will disagree with me: *We cannot master each of the firefighter disciplines listed above.* However, we can train to become proficient or "above average" in the disciplines that are needed at our respective fire departments. Without a doubt, each of us have particular interests in one or two specific disciplines. It is in these "areas of interest" that we must strive to achieve a mastery level. That is to say, we must develop our skills, knowledge, and abilities in these areas so that we are known as the "go-to" firefighters for those particular disciplines.

Some fire departments are fortunate enough to dedicate specific members and apparatus to specific disciplines, but this is rare. For example, a larger fire department may have the resources to assign firefighters who are solely dedicated to a rescue squad, or firefighters who only perform aircraft rescue and firefighting. In the same vein, certain fire departments may require their firefighters to be swiftwater technicians, ice rescue technicians, divers, and boat operators if bodies of water are in their response area. Or perhaps a fire station is near manufacturing or industrial plants that use chemicals—these firefighters typically have extensive training and knowledge in all things hazmat.

Take a moment to think about the high-hazard areas or occupancies in your respective fire department's response. Which specific fire disciplines are required?

5 FUNDAMENTALS TO BECOMING A WELL-ROUNDED FIREFIGHTER

We discussed the importance of training and education in *Chapter 8 - Competent*. Here are five fundamentals on how we can develop ourselves into well-rounded firefighters.

> "When we invest in our ourselves and our training, we will receive dividends for our entire careers."

1. OBTAIN AT LEAST ONE TRAINING CERTIFICATION OR SPECIALIZATION PER YEAR.

After graduating from the fire academy, the first specialization I obtained was my Hazmat Technician, then the next year I obtained my Fire Officer I. The following year I took Fire Instructor I, then Rope Rescue Technician, Fire Officer II, Fire Instructor II, Peer Fitness Trainer, etc. It takes time to attend such classes, and it may cost us money. But it is all worth it. We must invest in ourselves and our training to pay long-term dividends for our careers. After we accumulate training and certifications, let's not forget to attend applicable "refresher" classes to stay current in each discipline.

2. SPEND 15 MINUTES EACH SHIFT REVIEWING OUR RESPECTIVE DISCIPLINES.

For example, we should spend 15 minutes per shift reviewing ropes, knots, and raising/lowering systems if we are rope rescue technicians. Or we can spend 15 minutes per shift reviewing medication doses, rhythm strips, and EMS protocols if we are emergency medical technicians. Let's spend 15 minutes per shift reviewing forcible entry, ladders, search and ventilation if we are firefighters assigned to a truck company or we typically perform truck company operations. The options are limitless, and it takes only a relatively small amount of discipline to train for only 15 minutes per shift on each discipline. Doing so will keep our skills sharp and our knowledge fresh.

3. ALWAYS SEEK OUT MULTIPLE PERSPECTIVES.

One of the most dangerous ways of thinking is to be complacent with only our own fire department's training program. I have immense respect for training officers and training divisions, but there is only so much that they can cover. So much great information and resources exist outside the four walls of our own fire departments. We can attend local, regional, or national conferences, read books, listen to podcasts, watch online training videos, join an organization like the International Society of Fire Service Instructors, soak up the wisdom of mentors, etc. The options are truly infinite.

4. DO AWAY WITH THE "ENGINE VS. TRUCK" MINDSET.

Yes, it is comical to listen to die-hard "engine guys" explain why their job is more important than that of the "truckies," and vice versa. But if we are part of the other 90% of firefighters in the United States who are required to know and perform both roles, we must be proficient at everything. An "engine guy" who knows how to force a door, throw ladders, and vent a roof is a comprehensive firefighter. A "truckie" who knows how to simultaneously flow water and advance, as well as moving a hose line up stairs and around corners is a comprehensive firefighter. Such well-rounded firefighters are assets because they make the team stronger and more effective.

5. LEARN AS MUCH AS WE CAN ABOUT POSITIONS OTHER THAN OUR OWN.

It is understandable that not everyone wants to be an officer, chief officer, or a driver. But by learning other positions and their duties, we will become better overall firefighters and only get better at our current role(s). An additional benefit of knowing other positions' roles is that we can understand and empathize with those team members when they are actively completing a task. I have witnessed too many firefighters complain about other individuals in other positions who are *"not doing their job"* or *"their job is so easy."* Unfortunately, their ignorance is not allowing them to see the full picture of what each team member does. If we understand each others' responsibilities, we tend to appreciate them more and cut them a little slack, when needed.

> **"There is no time like the present to prepare for your future."**

Lastly, we never know when we will be presented with the opportunity to "go for the promotion." For example, if we are a firefighter who aspires to be an officer or driver, we must always be ready for the promotional process when it becomes available. Therefore, we should not start studying for the promotion when the upcoming promotional exam is announced.

10 KEYS TO BEING A THOROUGH FIREFIGHTER

In addition to being well-rounded firefighters, we will be *thorough* in everything we do. That is to say, we will pay attention to the details and sweat the small stuff. We will be meticulous and take great pride in every single aspect of the job.

There are countless ways how successful firefighters are thorough. Here are 10 keys to becoming a more thorough firefighter.

1. TRUCK CHECKS

Inspect every square inch of the truck: its equipment, tools, hoselines, cab, engine compartment, tires, etc. Every time we are at the firehouse, we must ensure everything is in working order, all is where it is supposed to be, and, most importantly, how to use every piece of equipment on our trucks. We also know multiple uses (if not dozens) for every tool on our trucks. For example, did you know that there are over 50 different ways to use the Halligan bar? It's not just for conventional forcible entry. Let's be students of our craft, and learn something new every time we are at the firehouse.

2. ORGANIZATION

It is a simple fact that those who are better organized are better prepared to execute. In all things, the details matter. Consider the way we set up our turnout gear at the fire truck before a call comes in. *Is it systematic, organized, and thorough? Or is it disheveled and chaotic?* The former leads to quick turnout times and the latter causes delays and a feeling of unease. Let's set up our gear the same way, every single time we arrive at the station or get back from a call.

One of the best ways to stay organized is to make a list of *"to-dos."* We can prioritize them based on their level of importance. Personally, I keep a revolving list on my phone in the "Notes" section. I keep the more urgent, short-term items at the top of the list, and the long-term items at the bottom of the list. If they have due dates, I put them with each *to-do*. After I complete an item, I delete it from the list. Also, I have found great success with setting reminders in my phone for items with specific dates and times. I even have recurring daily reminders, weekly reminders, monthly reminders, etc.

Lastly, as firefighters we are required to get and maintain a vast array of training certifications. Some of them have expiration dates, which require us to renew them periodically. Thorough firefighters organize these certifications so that they never lapse. The very moment we receive a new certification (or its renewal), the easiest thing to do is take a photo of it with our phones or scan it into "the cloud" (i.e. storage server). We can put it into a folder labeled "professional certifications." Again, we can take the extra step to set up a reminder in our phones for when we need to renew them.

3. CLEANLINESS AND MAINTENANCE

Whether it is the fire truck, the station, or our personal appearance, we as successful firefighters clean and maintain everything with honor and pride. Our saws are full of fuel and sharp, our hoseloads are packed uniformly to ensure rapid and efficient deployment, and our hand tools are free of debris and rust.

In the station, we clean, sweep, dust, and vacuum with the ultimate sense of ownership, knowing we are stewards of what has been entrusted to us. As for ourselves, we are well-groomed, physically fit, and prepared for any emergency that comes our way.

4. SIZE-UP

Whether we are the officer, driver, or firefighter, it is all of our responsibilities to continuously size up emergency scene conditions. If anyone observes hazards that are present, we will communicate them to others on the fireground. If we are the first-arriving officer, we deliver our initial incident report over the radio by communicating *"what we've got, where it's going, and how we plan to stop it"* (Chief Jim Silvernail - Kirkwood Fire Department, MO). We will be thorough, relaying directives to our crew and to incoming units, leaving no ambiguity on the strategy, tactics, conditions, actions, and needs.

5. COMMUNICATION

As we will discuss in *Chapter 14 - Communication*, thorough and open communication are absolutely critical when it comes to building trust, teamwork, and accomplishing the mission. Successful firefighters communicate thoroughly and frequently to minimize ambiguity and confusion.

If we are the company officer, we will communicate our expectations, daily agenda, training, constructive feedback, etc., to the crew. Let's not be afraid to overcommunicate, since *lack of communication* is one of the biggest complaints of nearly every firefighter.

6. FOLLOWING UP

Tied closely to communication, following up with others after making a request is a dependable way to make sure the job gets done. We just need to make sure we don't make too many follow-ups, or it will come off as nagging, controlling, or micromanaging.

Perhaps a fellow firefighter suffered a tragedy in their lives recently and is going through a hard time (e.g. death in the family, illness, divorce, etc.). A simple follow-up text message or call to ask them how they are doing goes a long way to let them know that we care. If the matter is more serious, we can get together with other firefighters and send a card and flowers to the firefighter and their family. Such a display of compassion will mean a lot to the firefighter who experienced the loss.

7. DOCUMENTATION

Although it may be our least favorite thing to do, we will be thorough with all of our documentation—*one of the necessary evils in our profession*. Whether it be incident reports, fire investigation reports, EMS patient care reports, workplace forms, disciplinary documentation, etc., we must *"document, document, document."* Thorough and accurate documentation protects us from future liability when it comes to potential lawsuits against the fire department and us as individuals.

8. PROBLEM SOLVING

Successful firefighters are problem solvers. When presented with any challenge or obstacle, we examine it in detail, break it down, and then come up with *plan A, plan B,* and *plan C* to fix it. We do not stop if we hit a roadblock, because we have the persistence, ingenuity, and teamwork to get any job done.

When it comes to fire extinguishment, thorough firefighters do not take salvage and overhaul lightly. We leave no corner unturned to make sure that *all* of the fire is *all* the way out. No fire company wants to hear the R word: *rekindle*. None of us wants to learn our fire wasn't all the way extinguished, and it spread because of our incomplete overhaul efforts.

> **"There is no such thing as a rekindle—it is merely a fire that was never fully extinguished."**

9. FIRE INVESTIGATION

If we are the fire investigator responsible for investigating the cause and origin of a we already know the importance of being thorough. Like overhaul, we will leave no stone unturned in our investigations, searching out every possible clue. One minute detail could be the difference between the classification of "cause undetermined" or "arson."

10. PATIENT CARE

For those of us who also deliver emergency medical services, we will perform a detailed patient assessment for every patient. Then we will administer the appropriate treatment plan within our scope of practice. Our differential diagnoses will be accurate, and our hospital radio reports always paint the full picture of the patient's condition for the receiving staff. We will provide the highest level of care and positive outcomes for our patients.

THERE ARE NO SHORTCUTS

George Washington Carver said it best: *"There is no shortcut to achievement. Life requires thorough preparation—veneer isn't worth anything."* As comprehensive firefighters, we will be well-rounded and thorough with everything we do. We will be successful because we take all aspects of the job seriously. We will not take shortcuts.

> "I don't have to train. I'm just going to hope for the best."
> -Unsuccessful Firefighter

ACTION STEPS

1. Do not settle for mediocrity. The next time a fellow firefighter chooses the easy way out, choose work.

2. What are your training specializations? Spend 15 minutes training on each the next time you are at the firehouse.

3. What is the next training certification or specialization that you want to obtain? Search for the next available offering and sign up for it today. If it is not immediately available, do a weekly search for it.

4. Do you lack organization? Get an organizer or calendar (it can be the digital or paper variety). Keep track of all your appointments, schedule, to-dos, etc. Look ahead to your schedule for the next week. In the evenings, look ahead to tomorrow's schedule.

CONSISTENCY IS KEY TO SUCCESS.

FirefighterSuccessBook.com

CHAPTER 10
CONSISTENCY

The word *consistency* has a double meaning. On one hand, it is the regularity, constancy, and steadfastness of the level of someone's performance. That is to say, a firefighter with consistency will have little variation in their habits and routines. Their sustained effort, and repetition of specific actions and behaviors over a long period of time are what help them repeatedly perform and execute at the highest level.

Everyone in the fire department knows which firefighters are consistent. They also know these are the firefighters who are dependable. They are revered as the "go-to" firefighters. We want to work with them. We want them on our crew. We want to fight fire with them.

> "Success is neither magical nor mysterious. Success is the natural consequence of consistently applying basic fundamentals."
> -Jim Rohn

Looking at consistency from a different perspective, it can also refer to the uniformity and quality of a material. For example, for precious metals like gold, their appearance and value are dependent on their purity and consistency. For diamonds and other precious gemstones, their clarity, beauty, and value are dependent on the absence of defects and irregularities.

From the inside out, our mindset, character, behavior, and actions must be consistent and compatible with each other. For example, a firefighter who says *"Training is the most important aspect of the job"* must reflect this statement with their actions. In other words, their beliefs, words, and actions will be consistent with each other. They must practice what they preach—they must walk the talk. If they do not, they will have a very difficult road to achieving success as a firefighter.

When it comes to leadership, firefighters want leaders who are consistent and fair. They want to know what to expect from them on a daily basis. They also deserve to have leaders communicate and enforce reasonable expectations for job performance. Leaders with consistency attempt to treat everyone with fairness. If a company officer has both a 30-year veteran or a 30-day rookie on their crew, both are treated with equal respect—*there is no favoritism or double standards.*

THE IMPORTANCE OF BUILDING SUCCESSFUL HABITS

Humans are creatures of habit. Habits are the actions we do repeatedly, typically in a way that requires little to no conscious thought. Some habits are constructive, and, unfortunately, some are destructive. The questions we must ask ourselves:

- What do we want to "build?"
- What are our goals?
- Are our habits paving the road to success? Or are they constantly taking us on detours?

Consistency and success are built by adopting and building more of the "good" habits, and doing away with the "bad" habits. Simply put, practicing good habits repeatedly (i.e. having a routine) yields a greater probability of success. We will discuss routines more in depth later in this chapter.

It takes commitment and intentionality to start tipping the scales in the right direction. Then it takes self-discipline, determination, and persistence to keep us consistent.

As we discussed in the previous chapter, consider the way we as firefighters set up our turnout gear at the truck before a call comes in. Successful firefighters will have it organized the same exact way every single time. This may mean our boots and pants are on the ground with our suspenders and hood positioned a specific way. It may also mean that our radio and radio strap are hanging on a certain hook, so that we can put it on in a certain order. We may choose to keep our SCBA facepiece preconnected to the regulator to save time. Whichever way we choose to set up our gear, we must stay consistent so that we build muscle memory, order, and efficiency.

By contrast, an unsuccessful firefighter will be disorganized and sloppy with their PPE. They will haphazardly throw items of their ensemble in random locations—simply hoping for the best. When the tones drop and they need to gear up, they are typically inefficient and slow (while everyone else is waiting on them).

ACTIVITY: SUCCESSFUL AND UNSUCCESSFUL HABITS

Here is a list of successful habits and unsuccessful habits. As you go through the list, circle all of the habits that describe you. Take some time with this exercise and most importantly: *Be honest.*

SUCCESSFUL HABITS	UNSUCCESSFUL HABITS
Arriving early	Arriving late
Being kind	Being rude, bullying
Practicing gratitude	Complaining
Giving others the benefit of the doubt	Gossiping about others
Working hard	Being lazy
Maintaining your poise	Losing your temper
Practicing routines	Not having routines
Waking up early	Sleeping in, hitting the "snooze" button
Reading books, articles, etc., for the purpose of personal development	Too much "screen time" (e.g. television, internet, social media, video games, etc.)
Exercising daily	Being sedentary
Communicating candidly	Being passive aggressive
Following through	Being unreliable
Speaking when necessary	Speaking too much
Standing up for yourself	Letting others walk all over you
Maintaining eye contact	Avoiding eye contact
Being honest	Lying
Practicing balance and moderation	Overeating and drinking too much alcohol
Having priorities and using discernment	Saying yes to everything
Giving encouragement	Constantly criticizing
Dressing and looking professional	Being a slob
Practicing compassion	Judging others
Practicing humility	Bragging
Contributing to the team	Being a freeloader or "weak link"
Drinking plenty of water	Drinking excessive soda, alcohol, caffeine, energy drinks, etc.
Praying or meditating	Failing to pause and reflect
Making a list of to-dos	Procrastinating
Setting reminders	Always forgetting

CHAPTER 10

Which habits describe you?

Pick three successful habits you want to adopt or get better at doing. What are they?

For each of these successful habits, write down specific actions you can take over the next two weeks that will help you adopt them.

Pick three unsuccessful habits that you want to eliminate. What are they?

For each of these unsuccessful habits, write down specific actions you can take over the next two weeks that will help you adopt them.

Depending on the habit, it can take anywhere from two weeks to several months to make it part of our lifestyles. Remember, we cannot rely on our level of motivation, which can be a roller coaster with its ups and downs. There will be distractions and hurdles that we must overcome. Did we stumble and fall off the wagon? Let's brush it off, hold our heads high, and get back up. We will stay the course and not let anything stand in the way of achieving our goals.

Be intentional.
Be disciplined.
Be determined.
Be persistent.
Be relentless.

> "We must all suffer from one of two pains: the pain of discipline or the pain of regret. The difference is discipline weighs ounces while regret weighs tons."
> - Jim Rohn -

BREAKING BAD HABITS: A PERSONAL SUCCESS STORY

Some who are reading this book have also read *Firefighter Functional Fitness,* a book I wrote with Chief Dan Kerrigan. Since being published, I have unofficially been dubbed as a "fitness guy." I will be honest: Fitness was always important, but I never thought I would write a book about it. I never thought I would have the opportunity to speak and share our message at fire departments and conferences on a national level. I never thought I would have a book that would reach and impact firefighters in dozens of countries worldwide. If you would have known my eating habits 10 years prior, you would have never guessed I would write such a book.

Like most everyone else, I have always had a sweet tooth. One of my unsuccessful habits was I consumed too much sugar, primarily in the form of soda and ice cream. *I used to drink two sodas per day and eat ice cream after every dinner.* As I was turning 30, I noticed that I was starting to gain some weight and getting a belly. Knowing that obesity, diabetes, and heart disease were part of my family medical history, I decided it was time to make a change.

I made it my goal to quit soda in six months, and to cut down on eating ice cream to only once per week. I went from drinking two sodas per day to only drinking one. I did this for a few weeks and started drinking a soda every other day. After a few weeks, I eventually switched to diet soda. Then I went to drinking one soda every few days, and finally I decided to quit completely. With the ice cream, I followed roughly the same process—taking small, incremental steps to achieve a larger goal. Over the course of a year, the results spoke for themselves: *I felt better about myself, I started drinking more water, and I lost 15 pounds!*

Consistency is what helped me achieve my goals. I turned my unsuccessful habit of consuming too much sugar around. With self-discipline and persistence, I achieved success.

> "Success isn't about talent or luck. It's about consistency."
> -Angela Kunschmann

ROUTINES: A RECIPE FOR SUCCESS

Adopting a regular routine is absolutely essential for building consistency and success. A routine is a sequence of actions (i.e. "habits in action") that we follow and regularly repeat. Establishing and following a routine requires a strong mindset and self-discipline.

With the desirable habits that we aren't yet doing, we must deliberately take action and then repeat them—*even if it is difficult and we do not want to do so*. This conscious and intentional effort ends up building muscle memory. Our brain's neural pathways start to fire together, creating a "mind and body" connection. It becomes easier to not only do the things we know we should be doing, but we do them with greater concentration, efficiency, and consistency.

Let's consider adopting the following habits into our daily routines:

- Shower and shave before going to bed.
- If we have to work tomorrow, let's prepare our uniform before going to bed.
- Go to bed at a decent time.
- Wake up early.
- Show up to the fire station early. *"Early is on time. On time is late. Late is unacceptable."* -Anonymous
- When we arrive, let's get our gear on the truck right away. Then let's check our trucks and equipment.
- Train, read, and exercise every day.
- Pray, meditate, or have a time of quiet reflection every day.

What habits would *you* add to the list? What will help you build consistency?

How you create and structure your daily routine is up to you. We must remember that although routines are incredibly important, we must also balance the ability to be flexible throughout our day and during the week. We cannot turn ourselves into robots just to become "slaves to routine."

As successful firefighters, we are adaptable and able to overcome unforeseen obstacles and conflicts.

> "Improvise, adapt, and overcome."
> -U.S. Marines

THE IMPORTANCE OF PACE

In our "instant gratification culture," society wants everything right now (without having to work for it). People want a magic pill that will help them lose 100 pounds in two weeks, yet it took them decades to put on that extra weight.

As it pertains to any worthwhile long-term goal, consistency and discipline make us work hard now for a reward that will come *later*. Being consistent with our habits and routines will help us make daily improvements that will improve our efficiency and effectiveness.

With long-term goals, people who practice deliberate, moderate effort on a daily basis will have a greater probability of success over those who have random, sporadic outpourings of maximal effort. Consider the novice endurance athlete and their training regimen. To successfully compete in their first competition, they must train consistently in a way that promotes daily improvement. Pace and consistency during their training play huge roles in their race day performance. By contrast, someone who has a completely random training schedule that lacks consistency will perform poorly and may not even complete the race.

> "Consistency in training yields consistency on the fireground."

As firefighters, we must do the same. We cannot sprint 100% of our time in the fire service, nor should we only sprint when we see the opportunity for promotion. We must keep a strong, consistent, and realistic pace that will foster growth and success with our goals.

And if we feel like we are stuck and not making progress, let's re-evaluate and keep moving forward. We cannot underestimate the importance of forward momentum, because inertia breeds inertia. Remember: *Progress is still progress, no matter how slow we are moving.* Stay optimistic—a breakthrough could be waiting just around the corner.

SHOW UP!

In life, we cannot win if we do not play. History is not made by those who stand on the sidelines (or sit in the station recliners all day). Success is reserved for those who show up and do the work. Let's *show up* every opportunity we can. And when we show up, let's make the most of every opportunity. Whether it is a training, a meeting, the promotional exam, or even being with our families, we will show up and give 100%.

> "The majority of success is rooted in just showing up."

CHAPTER 10

ACTION STEPS

1. Complete the "Successful and Unsuccessful Habits" exercise in this chapter.

2. The two best habits a firefighter can implement are waking up early and showing up early. Adopt these into your daily routine.

3. If it isn't already part of your fire station routine, set up your turnout gear on the fire truck the same way, every single time.

CONSISTENCY BUILDS CHARACTER AND SHARPENS THE MIND. CONSISTENT FIREFIGHTERS ARE TRIUMPHANT, WITH AN UNYIELDING INNER DRIVE.

— TONY FAHKRY —
(PARAPHRASED)

IF IT WERE EASY, EVERYONE WOULD DO IT.

FirefighterSuccessBook.com

CHAPTER 11
CHARACTER

We could be the most-skilled, most-knowledgeable, and most-experienced firefighters—but none of these will matter if we lack character. Character is truly one of the most important qualities that will contribute to our short-term and long-term success.

Character is comprised of many attributes: *resilience, integrity, honor, respect,* and *self-discipline* to name just a few. We will discuss these essential qualities, and explain why they are fundamental to our success.

RESILIENCE

In *Chapter 6 - Courageous*, we touched on character courage—an essential quality of successful firefighters. Possessing character courage is a requisite for becoming strong and resilient. Being resilient means we are able to use the challenges that confront us and convert them into opportunities for growth. It means we can use life's trials and adversity to become victorious. Steve Maraboli says it best: *"Life doesn't get easier or more forgiving; we get stronger and more resilient."*

> "The same boiling water that softens the potato hardens the egg. It's about what you are made of, not your circumstances."
> - Anonymous -

CHAPTER 11

Adversity builds character. History is filled with countless stories of extreme hardship, growth of character, and eventual triumph. Let's consider the following examples:

- **Franklin Roosevelt** was paralyzed from the waist-down by polio, yet he went on to be President of the United States four times.
- **Oprah Winfrey** was born into poverty, raped at age 9 and became pregnant at age 14, only to then experience the death of her infant son. Despite the odds, she built a media empire that would make her the richest black woman of the 20th century and the greatest black philanthropist in American history.
- **Frederick Douglas** was born into slavery and separated from his parents. Not only did he teach himself to read, he became the leader of the abolitionist movement with his groundbreaking antislavery speeches and writings.
- **Victor Frankl** was imprisoned in Nazi concentration camps and his entire family was killed by the Nazis in the Holocaust. Surviving the horror, he fled to America where he wrote *Man's Search for Meaning*, which has been named as one of the 10 most influential books in America.[13]

What do all of these historical figures have in common? They experienced overwhelming misfortune, yet they overcame it by being resilient and courageous. They are the epitome of character and success.

During our time in the fire service, we will be tested with struggles from every aspect of our lives—both on the job and in our personal lives. At work, we will be confronted with tremendous physical, mental, and emotional stress; with horrific incidents; loss of life; poor leadership; injuries; ridicule; gossip; and possibly not getting "the promotion." In our personal lives, we will have relationship or marriage challenges, child issues, health problems, economic stress, and may even wrestle with our faith.

> **"Life isn't easy. Life isn't fair. We can choose to be the victor, or we can choose to be the victim."**

When we are confronted with adversity, we have two options: give up or get back up. If we give up, we admit defeat. But when we get back up, we build character and become stronger for future battles. Adversity will undoubtedly come, but our strength lies in the choices we make to overcome it. Struggle, pain, and failure are the unavoidable necessities of life. We will embrace them and use them to build true strength and character.

---

[13] Spodek, Joshua. "12 Incrediblly Successful People Who Overcame Adversity." *Inc.* May 20, 2016, www.inc.com/joshua-spodek/12-incredible-people-who-succeeded-despite-adversity.html.

4 WAYS TO BECOME MORE RESILIENT

1. MAINTAIN PERSPECTIVE WITH FAILURE.

Did we forget to check a critical piece of equipment on our fire truck? Did we get into an accident while driving a fire department apparatus? Did we underperform during a hiring or promotional process? Although all of these things are

> "Failures don't stop you. Rejections don't stop you. Only you stops you."

important to our job performance and goals, we must keep everything in perspective. In *Chapter 5 - Committed*, we discussed the importance of establishing priorities. Although it is incredibly important, the fire service should not be our first priority; rather, our family and faith (if applicable to you) must come first. If our shortcomings at work do not detrimentally affect these two things, we should not give them undue precedence or attention. *Being a firefighter is important, but it is not everything.*

2. DO NOT DWELL ON FAILURE.

We must remember failure is neither final nor fatal. Everyone fails, falls short, and misses the mark. It may be deflating, embarrassing, and extremely difficult, but we must always see failure as an opportunity to learn, grow, and improve. The next time we fail and we are beating ourselves up, let's remind ourselves: *"Failure isn't final."*

3. PUSH YOURSELF PHYSICALLY AND MENTALLY.

Do hard things. When it comes to the training ground or the gym, let's regularly push ourselves to the limits. Let's test our preconceived physical and mental boundaries. Just when we think we are ready to throw in the towel or stop, we will do one more set, or one more drill. When we embrace physical discomfort, we can use it to build the unstoppable mindset. Building our physical toughness directly transfers over to building mental toughness and resilience—both go hand-in-hand. In the future, when we face physical, mental, or emotional adversity, we can draw from our previous experiences, finding the will to succeed.

4. REACH OUT FOR ENCOURAGEMENT.

Being resilient is not the same as being completely self-reliant, especially when it comes to overcoming adversity. If we believe we can always triumph over life's hardships all on

> "If it doesn't challenge you it won't change you."

our own, we are greatly mistaken. When we are struggling, one of the most dangerous things we can possibly do is isolate ourselves. If we are depressed and beating ourselves up for something we did, it is imperative that we reach out to those we trust and love. They will guide and encourage us through the dark times. And if they cannot provide the help we need, we must take the next step and have the courage to seek professional counseling. Getting help is not a sign of weakness; it is a sign of strength and character.

INTEGRITY

"Do the right thing" is a catchphrase that is thrown around flippantly in the fire service. But those of us with integrity actually live it out on a daily basis. We do the right thing because it is the right thing to do. We do the right thing when no one else is watching us. Most importantly, we do the right thing in spite of when the popular choice is the wrong choice. When the easy, popular choice might be laziness and mediocrity (i.e. lying in the station recliners all day), we will be training, reading, or working out—aiming to improve ourselves every day. This is "doing the right thing."

> **"Integrity never goes out of style."**
> -Jim George

Honesty is another essential quality of integrity. As firefighters with integrity, we value the truth, because we know the value of keeping our word. Successful firefighters know nothing will kill one's character and credibility faster than lying. We will not lie for the purpose of self-preservation, and we definitely will not lie about others to make them look bad.

Since we have integrity, we will not make excuses for our mistakes. When we mess up, we fess up. Since we are honest, we admit to our mistakes and own them, apologize, and then move on. Let's remember character and humility are inseparable.

> **"The most successful firefighters are those who value integrity above personal gain."**

Firefighters with integrity will also use whatever they possess to help others. *We do this out of a sense of duty to leave the fire service better than we found it.* Whether it is our knowledge, experience, rank, or resources, we will make it our goal to contribute to others and make the team better. We look for opportunities to serve fellow firefighters and those in our community. We don't do so for our own benefit, because having integrity means we do not expect anything in return.

Fire department leaders, do you use your position of influence to make others better? Do you welcome and genuinely listen to your firefighters' input? Do you give everyone the benefit of the doubt? Or do you put them off and disregard their ideas because you think that they are just "whining and complaining?"

Let's always remember that leadership is a privilege—it is an honor to be in a position where we can positively influence those under our command. Leaders who strive to serve others are the embodiment of character and integrity.

HONOR

As firefighters with character, we will honor our oaths to protect and serve our citizens to the best of our abilities. As we also discussed in *Chapter 5 - Committed*, a successful firefighter will honor the profession, its traditions, those who paved the way for us, and those who have sacrificed their lives while fulfilling their calling.

Because of the honor we possess, we will have unapologetically high standards. We will demonstrate excellence through everything: our training, our service delivery, public relations and education, and even our physical appearance.

A firefighter with honor will be of high moral and ethical character, being a positive representation of all other firefighters in the profession. If we are wearing a firefighter t-shirt when we are off duty and in public, we must remember we are representing everyone else in the profession. Whether we do something positive or negative, people will pay attention and our actions will reflect on all other firefighters. Are we getting drunk at bars or getting in fights? Are we driving erratically, cutting other drivers off (with our firefighter bumper stickers or plates)? Remember, if we "wear the t-shirt," we represent something much greater than just ourselves. We represent every other firefighter in the fire service.

Since we are firefighters with honor and character, we will always follow our promises with action. When we say that we are going to do something, *we do it*. Whether the task is big or small, we always follow through and we do so in a timely manner—with purpose and intentionality. We are men and women "of our word."

> **"Whoever can be trusted with very little can also be trusted with much, and whoever is dishonest with very little will also be dishonest with much."**
> **-Luke 16:10**

RESPECT

Firefighters with true character know they must first earn respect. We know that respect is not freely given, and we never simply demand it. It doesn't matter how smart, talented, or experienced a firefighter may be—respect must always be earned. Bruce Lee said it best: *"Knowledge will give you power, but character respect."* Character is built with respect, and respect is built with character—the two are forever intertwined.

> "Respect yourself. Respect others. Respect the job."

Whether we are one week out of the academy or we are one week away from retirement, a successful firefighter gains respect through their actions, their words, and through a blend of humility and confidence. It can take years to earn other firefighters' respect, yet it can all be lost in a single moment. Avoid the following *top 10 character killers* at all costs.

TOP 10 CHARACTER KILLERS

1. Being dishonest
2. Being egotistical
3. Being disrespectful
4. Not taking ownership of mistakes
5. Blaming others
6. Being lazy
7. Being late
8. Being a gossip
9. Being a complainer
10. Not following through with your promises

While we are on the topic of respect, let's discuss how we respect others. It has been said our character is defined by how we treat those who can do nothing for us. How we treat the newest firefighter in our department says a lot about our character.

Chiefs, do you welcome your newest firefighters into the department? Do you make it a point to know their names? Firefighters and company officers, do we respect, teach, and

> "Character is the foundation of respect and credibility."

mentor our rookies? Or do we belittle, shame, and bully them? Are we patient with them when they make a mistake? Or do we ostracize them and make them feel inadequate?

Let's be firefighters with strong character. Let's give the respect that we want to receive in return, even when someone apparently "doesn't deserve it." As Dave Willis has shared, *"Show respect even to those who don't deserve it, not as a reflection of their character, but as a reflection of yours."*

SELF-DISCIPLINE

Character is undoubtedly built and maintained through self-discipline. It takes discipline to create and adhere to a daily routine. It takes discipline to wake up early and be the first one to the station. It takes discipline to train every day. It takes discipline to

PEOPLE ASK ME, "HOW DO I GET TOUGHER?"
BE TOUGHER.

"HOW CAN I WAKE UP EARLY IN THE MORNING?"
WAKE UP EARLY.

"HOW CAN I WORK OUT CONSISTENTLY EVERY DAY?"
WORK OUT CONSISTENTLY EVERY DAY.

"HOW CAN I STOP EATING SUGAR?"
STOP EATING SUGAR.

"HOW CAN I STOP MISSING THAT GIRL OR GUY OR WHOEVER BROKE UP WITH ME?"
STOP MISSING THEM.

YOU HAVE CONTROL OVER YOUR MIND.
YOU JUST HAVE TO ASSERT IT.

- JOCKO WILLINK -
DISCIPLINE EQUALS FREEDOM

FirefighterSuccessBook.com

> "Discipline always starts with mindset."

work out every day. It takes discipline to eat right. It also takes discipline to say "no" to character-damaging behaviors: *laziness, mediocrity, complaining, gossiping, etc.*

Discipline comes back to mindset. Always. It does not rely on our emotions, nor our moods, and definitely not our level of motivation. If our mind is fixed on achieving a goal, we will achieve it because of our tenacious level of self-discipline. Remember: Motivation will come and go, but discipline is forever. Choose discipline.

BUILD UNBREAKABLE CHARACTER

Resilience, integrity, honor, respect, and self-discipline—if we adopt and live out these qualities, we will become firefighters with unbreakable character.

So many of us tend to worry about our reputation, but I'm here to tell you that is a waste of time. As the great Coach John Wooden shared, *"Be more concerned with your character than your reputation, because your character is what you really are, while your reputation is merely what others think you are."*

In simple terms: *Let's do the right things for the right reasons, and the right people will be attracted to our character.*

ACTION STEPS

1. Think about a very specific time in your life in which you experienced adversity. It could be a time when you failed, experienced a major injury or illness, lost a loved one, etc. How did you respond to this event? If a similar event were to happen in the future, how would you react?

2. Who is someone in your fire department or life who has character? Use them as a role model for building your character.

3. When wearing a firefighter t-shirt (both on and off duty), be cognizant of your words, behavior, and actions; they represent all other firefighters.

4. When a new firefighter joins your fire department, get to know them and make them feel welcome. Give them the respect that you would like reciprocated.

IT IS ONLY THROUGH SHARED STRUGGLE AND HARDSHIP THAT WE CAN ACHIEVE MORE, DO MORE, AND BE MORE.

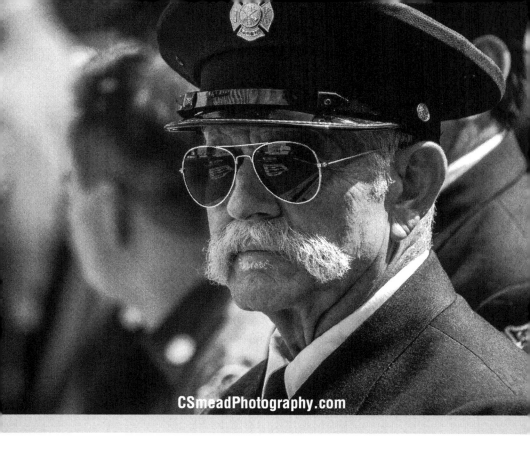

CREDIBILITY IS THE CURRENCY OF THE FIRE SERVICE.

FirefighterSuccessBook.com

CHAPTER 12
CREDIBILITY

Successful firefighters know credibility is the currency of the fire service.

Do our fellow firefighters and citizens believe us? Not only do they believe us but do they also believe *in* us? Do they find us to be reliable at the station, on the fireground, and outside of the fire department? In the simplest of terms, our credibility lies in whether or not others trust us. And to be considered credible, we must first be considered to be worthy of their trust.

Trust is built on honesty. *Do we tell the truth?*

Trust is built on integrity. *Do we do the right thing?*

Trust is built on action. *Do we do what we said we would do and do we lead by example?*

Trust is built on confidence. *Are we assertive, poised, and steadfast?*

Trust is built on competence. *Do we know our job?*

All of these qualities come together to lay the foundation for our credibility. And at the end of the day, our level of credibility determines our level of influence with others and the level of our success.

CHAPTER 12

WHY IS CREDIBILITY IMPORTANT?

Whether we are paid or volunteer, if we carry the title of "firefighter," we will live up to a higher standard. The public expects the highest level of service and professionalism from their firefighters, and we will deliver.

> "Before others believe in us, we must first believe in ourselves."

As public service professionals, our appearance and actions greatly impact our citizens' perception of us. Whether it is on duty, off duty, on a call, at a public relations/education assignment, and especially on social media, the way we carry ourselves always determines our credibility.

- Are we unshaven and overweight?
- Are our shirts untucked and our hats on backwards?
- Do we constantly say obscenities?
- Are we rude?
- Do we have a bad attitude at public relations events?
- Do we tailgate other cars or cut drivers off when driving fire department apparatus?

On the other hand:

- Are we well-groomed and physically fit?
- Do we have a welcoming attitude with everyone we meet?
- Do we go the extra mile to serve our public, even if it isn't necessarily part of our "job description"?
- Most of all, do we carry ourselves with pride and ownership, both inside and outside of the fire station?

As successful firefighters, we know that everything we say and do affects our credibility—*so we set the highest standard and we uphold it.*

> "Whether on or off duty, we always carry the title of 'firefighter.' Act accordingly."

CREDIBILITY IN LEADERSHIP

VOICES OF EXPERIENCE

Dan Kerrigan
Fire Chief - Upper Providence
Township Fire & Emergency Services (PA)
Co-author & cofounder of
Firefighter Functional Fitness

I am reminded of a simple post I saw on social media that included a picture of me as a very young and inexperienced assistant fire chief from several years ago. What I thought would result in a few days of good-natured ribbing from my friends ended up helping me reflect on just how powerful an example we all set. When the comments on the photo started to roll in, there was a sense of closure on my part. But, there was also clear validation that the way I carried myself 30 years ago had a positive and lasting impact on the firefighters I worked with. I had, in a certain sense, come full circle.

Jim speaks of several game-changing strengths in this book—essential qualities such as *curiosity, coachability, the champion mindset, charisma, and character.* Each is developed over time with the proper mindset and good mentoring. But what strikes me most is that the more we are willing to listen and learn from our mentors and peers, the more we are willing to accept we actually might not know it all and the better *our* example will be. We must never forget that every day we have an opportunity to mold someone—in many cases without ever saying a word to them.

> **"We do not get to decide *if* we are going to set an example in the fire service—we are being watched every shift, every day. What we *do* get to decide is the type of example we set."**

When I was a young firefighter, my passion for the job had occasion to overtake my actual level of *knowledge* and *experience*. In other words, I put my foot in my mouth a time or two, but who's counting? Thankfully for me, that phase was short-lived, because I had mentors who expected more from me. They gave me a chance to set a "course correction." I observed how these successful firefighters carried themselves, and how they displayed their attitudes, work ethic, and gratitude for being a part of our great craft. More than any training session I ever attended, learning from their example was the most valuable lesson I could have ever received.

Now, as a fire chief of a young, eager, and growing department, there isn't a day that goes by that I do not think about the importance of the example I set. As a young firefighter, I had no idea of the impact I was having on those around me. I am forever thankful there were others who did.

One of the best ways to give back to the fire service is to set the right example for those who are coming behind us. Each of us has a responsibility in this fire service circle of life. In this book, Jim has outlined a road map to success for your personal fire service journey. Take the lessons taught in this book seriously. Apply the principles, and set the right example.

Allow me to add to Chief Kerrigan's message. If we want to have true influence on our firefighters and our fire department, we must be credible. No one will follow a leader (formal or informal) if they lack credibility, because no one trusts them. Our ability to lead and gain others' trust comes from our competence, confidence, and character with our people.

As Dan described, the quintessential way to build trust with our people is to always lead by example—every single day and in every single way. We must remember that others will follow our example, not our advice. They will remember our actions, not necessarily our words.

> "Example is not the main thing in influencing others. It is the only thing."
> -Albert Schweitzer

If we give specific expectations to our followers, we must live them out 100% of the time. If we want a clean firehouse, we as leaders must participate in cleaning. If we want our crew to be well-trained and physically fit, we must train and work out. Credible leadership isn't pointing fingers and barking orders; it is *"knowing the way, going the way, and showing the way."* (John Maxwell)

5 WAYS TO BUILD CREDIBILITY

When we are new firefighters (and new officers), building credibility all comes down to proving ourselves. It doesn't happen in one shift—it is built by proving that we are reliable

> "Credibility cannot be bought, it is built."

and trustworthy, day in and day out. Here are five ways that we all can build credibility.

1. DO THE SMALL THINGS RIGHT.

As previously discussed, our foundation to success is built on mastering the basics. When we have the opportunity to do the "small things," let's do them right. Whether it is checking the truck, cleaning the station, reloading a hose line or the hundreds of other small things we do, we will prove ourselves credible when we do them right. As Chief Frank Viscuso has shared, *"If you can't be counted on to take out the garbage, wash apparatus, or complete reports on time, how do you expect to be considered reliable at a structure fire?"*

2. KEEP YOUR WORD—100% OF THE TIME.

Let's do what we said we would do. Let's be firefighters who follow through with our promises. Our "yes" means "yes," and our "no" means "no." Did we say that we were going to work out today? *Do it.* Did we say we were going to take a specific firefighter training certification this year? *Do it.* Did we say that we were going to take the next promotional exam? *Do it.* Let's not be firefighters who have their feet halfway in the pool—*let's be all in.*

3. BE ON TIME.

If there is one credibility builder that requires minimal effort and absolutely zero talent, it is showing up on time. Not only should we be on time but it benefits us to arrive early because we can mentally prepare for the shift, training, meeting, event, etc. Conversely, being consistently late is a guaranteed way to show we are unreliable, and it will ruin our credibility. So for every event we go to, let's prove ourselves to be reliable and credible by arriving early.

4. DEMONSTRATE TRUST IN OTHERS.

How can anyone trust us and find us credible if we do not first trust them? Leaders, if we give our firefighters a task, let them complete it the way they see fit. Let's let them know what we want the outcome to be, but not necessarily tell them how to do it. Above all, we cannot stand over their shoulder and micromanage them. It's okay if they don't do it exactly how we would do it. And if they make mistakes, that is okay, too. Let them be free to make their mistakes so that they can learn from them.

5. HONESTY IS ALWAYS THE BEST POLICY.

> "Credibility takes years to build, but can be lost in only a moment."

We will fully discuss the importance of honesty in *Chapter 15 - Candid*. But for now, we must be crystal clear on how dishonesty negatively impacts our credibility. We must never lie, cheat, or steal. Doing any one of these demonstrates a lack of character and integrity, and they will deplete our trust account with others. Once we lose others' trust, we lose credibility. Rebuilding both will take years to accomplish.

REPUTATION VS. CREDIBILITY

Too many firefighters are more concerned with their reputation, when they should focus on building their credibility. Reputation is what others believe *about* us. Credibility is whether or not others believe *in* us. Reputation can be based on an individual's perception of us—which can either be accurate or inaccurate. Credibility is based on others directly observing our words, actions, and the example we set. It's nice to have a good reputation, but that will only get us so far.

I was once asked by a new firefighter on social media how he could "make a name" for himself in the fire service. He desired to become well-known and build a national reputation. In other words, he wanted to become "fire service famous."

My response was simple: *"Don't focus on making a name for yourself. Forget about your reputation. Instead focus on doing the right things for the right reasons, and you will naturally gain others' trust and build credibility. Over time, if you are meant to have a certain level of recognition and influence in the fire service, it will happen naturally."*

The bottom line: Let's first focus on building our credibility, and a desirable reputation will follow.

IT ALL COMES BACK TO OUR CREDIBILITY

Everything shapes our credibility: *our words, attitudes, behaviors, habits, actions, inactions, and mistakes.* We can never be perfect, but we will do our best. When we make mistakes (and we most definitely will), we must demonstrate humility and ownership. As Mark Batterson preaches, *"When we mess up, we must fess up."*

Finally, let's remember that simply wearing the uniform doesn't make us successful firefighters—our credibility does. And credibility is built on our competence, confidence, and character.

ACTION STEPS

1. Identify a successful firefighter or person in your life who has a high level of credibility. What are the character traits that make them so credible?

2. Identify a firefighter or person in your life who lacks credibility. Why do they lack credibility?

3. Identify three things you will work on over the next month to bolster your credibility as a firefighter. For example:
- Becoming more trustworthy
- Training more or becoming a more competent firefighter
- Arriving early to your shift, training, and events
- Leading by example
- Fulfilling your promises
- Taking action

SUCCESSFUL FIREFIGHTERS USE THEIR CHARISMA TO POSITIVELY INFLUENCE EVERYONE AND EVERYTHING AROUND THEM.

FirefighterSuccessBook.com

CHAPTER 13
CHARISMA

Charisma is one of those intangible, sought-after virtues that every successful firefighter must possess. It may be difficult to describe, yet it is made up of several desirable qualities: *passion, positivity, enthusiasm, warmth, and confidence.* When we combine all of the qualities into a singular purpose, charisma is best used to positively influence others. In the fire service, charisma is absolutely essential to strong leadership. Good leaders use their charisma to inspire and persuade those under their leadership.

> "Know your job, set a good example for the people under you and put results over politics. That's all the charisma you'll really need to succeed."
> -Dyan Machan

For example, we have all been taught by fire service instructors who have charisma, and by those who have none. Those with charisma are relatable, energetic, and passionate about what they are teaching. They use humor and excitement, as well as interesting stories and analogies. They capture our attention and make us want to learn more. On the other hand, the instructors who lack charisma are monotone, boring, and unrelatable. Unfortunately, they extinguish our desire to learn and train.

In our profession, we constantly interact with people from all types of backgrounds, ethnicities, personalities, socioeconomic status, etc. Regardless of who we are serving, we must be approachable, relatable, and even persuasive at times. In other words, a successful firefighter must become a "people person." For some this may come naturally, and for others it may not.

When I first joined the fire service, I will honestly admit that I was a through-and-through introvert. But as I went on more calls, interacted with the public more, and interacted with my crew members at the firehouse, I slowly grew into being a people person. This process took a lot of effort on my part because it required me to get out of my comfort zone. I made myself hang out at the kitchen table more, go to more off-duty social events, and I found extra opportunities to interact with the public during public relations activities. I am pleased with how I have grown, especially since I am now a company officer, instructor, public speaker, and leader within my organization.

> "Charisma has nothing to do with energy; it comes from a clarity of WHY. It comes from absolute conviction in an ideal bigger than oneself."
> -Simon Sinek

CHARISMA: THE GOOD, THE BAD, AND THE UGLY

Possessing charisma is a desirable trait. However, we must temper it with humility and noble purpose. Throughout history, there have been influential leaders who have used their charisma for selfish ends: Personal gain, prestige, power, and self-gratification. If we are to possess charisma, we must always use it for good.

Charisma for the sake of charisma means nothing. If we use our charisma to say *"Look at me!,"* we are missing the mark. There are plenty of firefighters who are good storytellers. Yes, it may be fun for them to have everyone hanging on their every word, laughing and enjoying their stories. *Consider these questions: Are we using our charisma for good? Are we using it to inspire passion and confidence in those around us? Or are we merely using it to entertain others or be the center of attention?*

> "How can you have charisma? Be more concerned about making others feel good about themselves than you are making them feel good about you."
> - Dan Reiland -

As successful firefighters, we must always use our charisma and influence to positively impact others. Under no circumstance will we use them to gossip about others or make them look bad in an attempt to make ourselves look good. I have seen too many firefighters and officers lose respect and credibility by using their platforms to tear others down. It becomes a lose-lose situation for everyone involved.

> "But charisma only wins people's attention. Once you have their attention, you have to have something to tell them."
> -Daniel Quinn

We must also know negativity is the enemy of charisma. Undoubtedly, every firefighter has been a victim of toxic leadership at some point in their career. Negative leadership kills every firefighter's passion and motivation. If we are to be firefighters with charisma, we must be cognizant of the messages we are sending—both verbally and nonverbally. Our attitudes and behaviors have a direct impact on how others view us. If we want to be seen in a positive light, we must stay positive.

As successful firefighters, we will use our charisma to inspire positive change in our fire departments and in our communities.

ACTION STEPS

1. Do you feel that you are lacking charisma? Identify people in your life who have it. In your eyes, what makes them charismatic? Adopt these qualities and use them in your interactions with others.

2. What fire service subject are you passionate about—*engine company operations, truck company operations, firefighter wellness?* As you progress in your career, develop this passion by learning as much as you can about it. Use your charisma to spread your passion and knowledge about it to others.

3. Smile more often. *Really.* People who smile more are more approachable, relatable, more confident, and are better leaders.

COMMUNICATION, THE HUMAN CONNECTION, IS THE KEY TO PERSONAL AND PROFESSIONAL SUCCESS.

- PAUL J. MEYER -

FirefighterSuccessBook.com

CHAPTER 14
COMMUNICATION

Communication is about human connection, and, ultimately, our profession is all about people. In particular, it is about serving people. In order to serve, we must be able to connect and interact with all different kinds of people on different levels. We communicate with our citizens, with our fellow firefighters, with our commanding officers and subordinates, with our elected officials (e.g. mayor, board of directors, city council, etc.), and especially with our friends and family.

Our ability to communicate effectively will have a large role in our personal and professional relationships, which will ultimately determine our level of personal and professional success. If we are unable to adequately express ourselves and our ideas, and unable to understand others, then we are holding ourselves back and limiting our potential.

We tend to think of communication as *what we say*, but it is so much more. Listening, our nonverbal gestures and facial expressions, our tone of voice, our emotional intelligence, and the written word all significantly impact how we express ourselves. They also affect how others interpret us and how others will express themselves to us.

Entire books have been written on communication, but we will devote this single chapter to discussing it in a way that specifically applies to firefighters. It will go much deeper than a discussion of the mundane communication model (e.g. *sender, message, medium, interpretation, and receiver*). We will discuss a comprehensive approach that will develop firefighters into effective communicators for a lifetime of professional and personal success.

CAN YOU HEAR ME NOW?

Most think of communication as effectively *sending* our message to others. But before we ever send or say anything to anyone, *we must first listen*. The foundation of communication is listening, and our first priority is to become the best listeners we can be.

When we were infants, we did not come out of the womb talking—*we listened*. And during this year or so of listening, we were learning. We first learned how to interpret our family's messages, then we processed them, and finally we started to formulate our own messages back to them—through gestures first and then spoken words.

> "The foundation of communication is listening."

Let's take it a step further and ask ourselves these questions: Are we only *hearing* what others are saying? Or are we actually *listening*?

Hearing implies the sender's words are making their way into our ears, but we aren't necessarily engaged with their message. As the cliché goes, their words could be "*going in one ear and out the other.*"

Listening, by contrast, means we are actively participating with their message. We are looking them in the eyes. We are processing what they are saying to us and thinking about it. We are minimizing distractions (especially our phones), and we are making sure they know that everything they are communicating is a priority. We show we are actively listening by rephrasing what they say, and we demonstrate our interest by asking them follow-up questions.

Being a good listener does not come naturally to most firefighters, especially since 93% of the U.S. fire service are men.[14] Trust me, I speak from my own experience of being a male and a recovering bad listener—*just ask my wife*.

Listening is a skill, and just like any other skill, it takes time and practice to develop. As a company officer, my crew members are frequently coming to me to either ask a question or to inform me that something needs attention. I may be on my phone sending an important text message or email when they start to communicate with me. But I have had to learn that I must stop what I am doing, put my phone in my pocket, look up and listen to them, and address their concern. Nothing is more disrespectful than when someone won't even look up from their phone to listen to what the other person has to say for a meager 10 seconds. We have all been on the receiving end of this treatment.

---

14 Evarts, Ben. and Stein, Gary P. U.S. Fire Department Profile. *National Fire Protection Association*, February 2020, www.nfpa.org/-/media/Files/News-and-Research/Fire-statistics-and-reports/Emergency-responders/osfdprofile.pdf.

> "Leaders who don't listen will eventually be surrounded by people who have nothing to say."
> - Andy Stanley -

Pledge to not be that individual. Let's have the self-discipline to put our phones away when someone is talking to us.

"BEING HEARD" AND DISAGREEMENTS

So many times, we are more worried about "being heard" first, as opposed to understanding what someone else is wanting to communicate. When we are too quick to speak, we never fully listen. Are we able to "hear" what *wasn't* said? As we shared previously, Dr. Stephen Covey said it best: *"Seek first to understand, then to be understood."* This principle is especially useful when it comes to mitigating conflict.

> "When we are quick to speak, we never fully listen."

When we let others express their perspective first and then truly attempt to understand where they are coming from, we are accomplishing two important tasks at once. First, we are getting the full story from that person, removing our bias, and potentially incorrect assumptions. More often than not, we need to gain perspective and see things through the other person's eyes. Second, we are displaying humility to that person and demonstrating to them that what they have to say has value. That act alone can remove a great deal of tension and break down defensive barriers, allowing for honest communication. When communication is honest and free, conflict is resolved much quicker.

How do we communicate when there is disagreement? Do we keep our composure and present our case with poise? Or do we shout, argue, or even try to belittle the other person? Are we willing to compromise for the greater good? Do we have the humility to sacrifice "being right" for the benefit of the relationship? What happens if the other person starts the conversation by yelling, making false accusations, and belittling us? Most likely we either shut down and communication stops altogether, or, even worse, *we do the same back to them.*

It is perfectly acceptable to be heard, to express our opinion, and to communicate our side of the story. And disagreements are okay, as long as we participate in them with humility, respect, and an open mind.

> "Don't raise your voice. Improve your argument."
> -Desmund Tutu

In the presence of conflict, there is also something to be said for someone who maintains their poise and integrity while someone else is shouting at them. When we do so, others will respect us for it and value what we have to say.

CHAPTER 14

> "Choose to be the firefighter who keeps their cool when others are losing theirs."

TRUST

Have you ever been under leadership that you did not trust? How easy was it to communicate with those who were in charge? Or have you had leadership that refused to communicate fire department information because they used it as a display of power? Or perhaps you have worked under a senior firefighter or officer who were "knowledge hoarders" and refused to share what they knew. In all of these circumstances, you most likely were unwilling to share your perspective because you knew that your input would not be valued or even received.

> "Communication requires trust. Trust requires communication."

A lack of trust begets a lack of confidence. A lack of confidence yields a lack of communication. And a lack of communication leads to a lack of morale. Without trust, confidence, communication, and morale, people will not be honest about how they feel, and they will not express what they really want.

On the other hand, all of us have had relationships in our lives where we have trusted the other person 100%. Because of this trust, we were able to freely communicate *anything* to that individual. We were able to be vulnerable and share our honest opinions and true feelings. We were also able to hold these same individuals accountable and tell them hard truths because we had built mutual trust.

Building a strong team requires trust between every member. The strongest teams know each member will do

> "When trust is high, communication is high. When trust is lacking, communication is lacking."

whatever it takes to protect each other. Strong team members are willing to go to extraordinary lengths to sacrifice themselves for the good of the team and the mission.

As hard as it may seem, leaders must ask their people for honest feedback. They can ask how things are going and what can improve. When a follower says something that hits hard, leaders must be humble, coachable, and teachable. When a leader shows they are willing to improve and then take action to do so, trust is built immediately.

> "Trust is not built overnight. It is cultivated over time with our actions, our character, by earning others' respect, by keeping our promises, and with open and frequent communication."

ADD VALUE, NOT JUST WORDS

When we speak, are we doing so just to fill the quiet space, or perhaps to make our every opinion known about every subject?

> "Wise men speak because they have something to say; fools because they have to say something."
> -Plato

By contrast, is what we are saying actually adding value to the conversation? In *Chapter 1 - Coachable,* we discussed the importance of humility, and using the THINK principle. Before we speak, is what we are about to say *True, Helpful, Inspiring, Necessary, and Kind?* In other words, do we really need to say it?

Some of the wisest firefighters I know are not the ones who speak a lot. Rather, they are the ones with the quiet confidence to speak only when necessary. They listen to the words others have spoken, take them in, process their thoughts, and then speak only as needed.

There is a reason why we have heard the phrase *"silence is golden"* countless times before. We all have witnessed the fool who talks too much. It seems they never know when to stop talking, and even after they have stumbled over their words they continue to dig themselves into a deeper hole. And let's be honest: *There have been times when we have been that very fool.*

When we talk, let's be sure to add value to the conversation, not just words.

BE CONCISE

Undoubtedly, we all know that one person who loves to hear their own voice, so they talk incessantly. They are the ones who get the eye rolls and sighs from the group that is listening.

We must focus on the *quality* of our words, not the *quantity*. One of the most important lessons I learned in life is that words have value, and we should not give them away freely. Our words are a reflection of who we are. They reflect our character and what we stand for. Let's choose them wisely, and choose to use them for maximum impact.

CHAPTER 14

There is great power in the ability to get our point across in one or two sentences, as opposed to someone who requires 10-20 sentences.

If we are instructors or trainers, let's keep it simple, short, and sweet. Tell firefighters *what* they need to know, *why* they need to know it, and *how* to do it or implement it. If we have to speak at a public education event or give a speech to an audience, we will be brief and to the point.

The Gettysburg Address was delivered by President Abraham Lincoln at the dedication of the Civil War's national cemetery on November 19, 1863. It is one of the most revered speeches in all of American history. Yet it is less than two minutes long and less than 300 words. Edward Everett, one of the best-known speakers of his time, was also at the ceremony with Lincoln. He too spoke, but his speech lasted two hours! The day after the event, Everett said the following to Lincoln: *"I wish that I could flatter myself that I had come as near to the central idea of the occasion in two hours as you did in two minutes."*[15]

Whether at a training, an event, or even at a gathering, let's be concise with our message. Everyone's time is valuable, and they will appreciate our brevity.

TONE AND TACT

While we are on the topic of being direct, let's remember the way we say something can mean everything when it comes to interpersonal communication. Our tone and tact will greatly influence our conversations, mainly how others interpret our message. Marriage has taught me a great deal about this.

At the firefighter level, the way we ask questions affects others' responses. If we have a rookie firefighter who constantly asks *"why"* with a curt attitude, they will have a very short fire service career. However, if they soften their tone of voice and use a little tact by saying: *"I'm not so sure I understand. Could you explain it a bit differently?"*, then their question will be received and answered with open arms.

For those of us in leadership roles, the tone and tact we use with our people will positively or negatively influence their morale. Do we rudely bark orders to our members (and our children or parents) like they are robots? Or do we use a warm tone and treat them like real people who have value? No one wants to work for an officer who verbally assaults their firefighters. Everyone wants to be respected, and it starts with the way we talk to others.

All of this does not mean we have to sugarcoat everything we say. However, we must be cognizant that the way we say something affects others' perception of our message. And as we all know, perception becomes reality.

---

15 "Gettysburg Address" *Encyclopedia Britannica*, 2019, www.britannica.com/event/Gettysburg-Address

CLOSED-LOOP COMMUNICATION

When lives are on the line, effective communication at emergency scenes is critical. There have been numerous firefighter injuries and even fatalities that have resulted from poor communication or no communication. Too many times firefighters have been at an active incident and someone gives a long, detailed message over the radio, to which the receiver simply says, *"That's clear."* Did the receiver really hear and understand the sender's message? Maybe, maybe not. We must do better.

A proven way to improve fireground communications is through closed-loop communications. In the simplest of terms, the sender relays their message to the receiver, then the receiver repeats what they heard (and understood) back to the sender, confirming the message. When performed over the radio, this may take up a bit more air time, but it undoubtedly cuts down on confusion.

Away from emergency scenes, interpersonal closed-loop communication is also very important. Whether we are communicating with someone in person, via text message, or via email, we want to know they received our message. Whether we are making a simple request or just passing on information, a simple *"okay"* or *"got it"* will typically provide enough acknowledgment. With my crew members, it is one of my basic expectations for them to respond as mentioned above if I send them an important text message or email.

If someone emails me a specific request for information or to complete a task for them, and I cannot complete it immediately, I will respond, *"I will take care of it and get back to you as soon as possible."* That way the sender knows that I received their request, it is a priority to me, and I will be working on it. Anytime a chief officer emails (or texts) me important information, I respond with *"Message received,"* to confirm that I have read it.

These processes may seem overkill, but I will tell you from personal experience: Nothing is more frustrating than sending an important, time-sensitive message to someone, they never respond, and then I follow up a week later to their response of *"Sorry, I never got your message."*

If you want to achieve success in your fire service career, use closed-loop communications as much as possible.

PUBLIC SPEAKING

We are in the people business. An essential part of our business is to communicate with our citizens, educating them on how to improve their safety, as well as educating them on what we do as firefighters. Call it *public education, public relations, or community risk reduction*—all of these require us as firefighters to be able to speak publicly. We give station tours, truck tours, and fire safety talks at schools frequently. There may also be a time when we are interviewed by the news media concerning an emergency incident.

> "You can have brilliant ideas, but if you can't get them across, your ideas won't get you anywhere."
> -Lee Iacocca

Most firefighters have a fear of public speaking, which is only natural for 99% of the population—*I know because I used to fear speaking in public.* Follow these six tips for improving your abilities to speak publicly.

1. TAKE A PUBLIC SPEAKING CLASS.

In college, one of the most important classes I took was *Speech 101*. Like almost everyone reading this, I hated public speaking. But taking this class made me get out of my comfort zone and grow. It gave me confidence to speak in front of my peers and other audiences.

2. ORGANIZE YOUR PRESENTATION.

Whether you are speaking to a large or small audience, make a simple outline of what you want to cover. Create simple, step-by-step lesson plans for station tours, truck tours, and school fire safety talks. The more that you give your message, the less you will need to refer to your outline.

3. "MASTER YOUR MESSAGE."

Nothing will give you more confidence to deliver your message than knowing your material front and back.

4. PRACTICE, PRACTICE, PRACTICE.

Rehearse your delivery as much as possible. Know all of the main points by memory, with the ability to improvise on the smaller details.

5. EVERY CHANCE YOU GET, SPEAK TO A GROUP.

Start with smaller groups (3-5 people), and then naturally expand your audience (10-20, then 30-50, etc.). Fire station tours are great for building your confidence with smaller groups.

6. REMEMBER THAT IT IS ALL ABOUT THE AUDIENCE.

Smile, have fun, and interact with them. Ask them questions and make them laugh. Share your message with a healthy blend of humility and confidence. Make them feel comfortable and connect with them on a personal level. They will undoubtedly remember your message.

Being able to speak in the public setting is important for all firefighters, especially as we progress up the ranks. The fire service adage *"You can't talk the fire out"* is true for the fireground. However, we can speak to our public and educate them on how to prevent fires in the first place. The more we speak in public, the better we will get. It may seem uncomfortable at first, but with practice, our confidence and abilities will grow.

PROFESSIONAL WRITING

Some may call it *formal writing*, but we will refer to it as *professional writing*. Most firefighters do not think about professional writing, but it is a vital aspect of our duties. We use professional writing for emails, alarm reports, incident narratives, vehicle accident reports, workers' compensation reports, proposals and purchase requests, and more.

> **"Either write something worth reading or do something worth writing."**
> **- Benjamin Franklin -**

Like it or not, spelling, grammar, punctuation, and phrasing matter. Not only are the quality of these items a reflection of our writing abilities but people who read our writing will also interpret them as a reflection of our intelligence. Unfortunately, if we have poor spelling and grammar, it will reflect poorly on our credibility and our message may not have the impact that we intended.

The quality of our writing is especially important when it comes to the legal process and lawyers. If we are ever called to testify in court as a witness, plaintiff, or even as a defendant, lawyers will read over our alarm reports with a fine tooth comb. They will look for misspellings, poor grammar and punctuation, and any inconsistencies. They will use these errors to lessen our credibility and associated testimony. Let's do ourselves a favor: We should have the mindset of writing every alarm report like it is going to be read in court.

One of my hobbies is woodworking, specifically finished carpentry. I love the saying *"Measure twice, cut once."* The same can be said for composing emails: *Read it twice, send it once.* And please do everyone a favor: If our email says there is an attachment (photo, document, etc.) *let's make sure we actually attach it*.

If we are unsure about the quality of our writing, let's have someone we trust proofread it before we send it. They will surely provide us with great tips that will benefit our writing for the rest of our careers. One last thing regarding emails: Let's never send an email when we are angry. It is okay to compose the email when we are frustrated, but let's wait to review it, edit it, and send it after we have calmed down. We cannot *unsend* an email once we click "send." Let's make sure we do not say anything that we will later regret.

10 TIPS FOR WRITING SUCCESS

Consider these 10 tips which will drastically improve your professional writing.

1. Take a basic composition and grammar course at a local college or online.
2. Enroll in a college fire science program to pursue an associate's or bachelor's degree. By writing dozens and dozens of essays and research papers, our writing knowledge, skills, and abilities will grow tremendously.
3. When writing a narrative for an alarm report or incident report, write the events in chronological order. This method will keep our writing on track and will be the easiest for the reader to follow.
4. In emails, address supervising officers with their title: *"Chief," "Captain," "Lieutenant,"* etc. At the very minimum, at least write *"Sir"* or *"Ma'am."*
5. Begin formal fire department requests, memos, and proposals with: *"To, From, Date, Regarding."* For example:
   - To: Fire Chief Smith
   - From: Captain Moss
   - Date: 1-1-2020
   - Re: Fitness Equipment Proposal - Station 1
6. End formal proposals and requests with *"Respectfully Submitted, [your name]"*.
7. Have an up-to-date resume, even if you do not plan on applying for a different job or going for a promotion.
8. Add a simple, custom signature to your emails. This can be easily set up by going to your email software's settings. Here is an example:
   - Jim Moss (name)
   - Captain (title)
   - Cell: 555-555-5555 (phone number)
   - SpringfieldFireDepartment.gov (website or email)
9. Know the difference between *"there," "they're,"* and *"their,"* as well as *"your"* and *"you're."* That's all I'm going to write about that.
10. Most importantly: After you are done writing anything, take an extra moment to re-read what you just wrote. Pay attention to spelling, grammar, punctuation, flow, phrasing, etc., and make the appropriate corrections.

20 RULES FOR SUCCESSFUL COMMUNICATION

To summarize what we have discussed in this chapter, use the following rules on a daily basis, throughout your career, and in your personal life to communicate successfully.

1. Listen more than you talk.
2. Think before you speak.
3. Understand others' perspective first before making assumptions.
4. When meeting someone for the first time, offer a smile and a firm handshake.
5. Maintain eye contact.
6. Be honest.
7. Be tactful.
8. Be concise.
9. Be candid.
10. Be nice.
11. Acknowledge that you heard what someone said to you.
12. When someone is talking to you, put your phone away.
13. Say *"please," "thank you," "you're welcome,"* and *"I'm sorry"* when appropriate.
14. Address commanding officers by their title: *Chief, Captain, Sir, Ma'am, etc.*
15. Use the chain of command—both up and down.
16. Proofread all written correspondence before sending it.
17. Give everyone the benefit of the doubt.
18. Don't gossip or complain.
19. During disagreements, always keep your poise and be respectful.
20. Never respond to an email when you are angry. It can wait.
21. BONUS: Before you post anything on social media, remember that what you share reflects on all firefighters. Represent the profession with honor.

Like public speaking and our firefighter skills, the only way to get better at communication is to get those "sets and reps" in. The more we speak publicly, and the more we write professionally, the better we will get at them. Whether it is a training, public relations event, email, alarm report, or formal proposal, let's make sure to give our best effort and represent ourselves well.

CHAPTER 14
ACTION STEPS

1. Put the phone down! Better yet, put it away. When everyone is at the firehouse kitchen table, talk. Don't become a "digital zombie," entranced by a little rectangular screen.

2. If you don't have one, create a resume. If you already have one, update it for your next career goal (e.g. promotion, applying at a different fire department, etc.).

3. If you have never written one, write a formal memo for a mock proposal. Use Tip #5 from *10 Tips for Writing Success*.

4. Start your very own Firefighter Journal. Document noteworthy calls, training, positive and negative events, etc. When it is time to retire, you will be amazed to go back and reminisce over your career.

5. If public speaking makes you uncomfortable, get out of your comfort zone. Be the next one to volunteer for the station tour, public relations or public education assignment.

BE SINCERE.
BE BRIEF.
BE SEATED.

— FRANKLIN D. ROOSEVELT —

LEAVE AN IMPRESSION WITH THE QUALITY OF YOUR MESSAGE, NOT THE QUANTITY OF YOUR WORDS.

FirefighterSuccessBook.com

CHAPTER 15
CANDID

We live in an overly sensitive society where most people fear the idea of speaking their minds. Instead of being honest about how they feel, they walk on eggshells, afraid they might offend someone. In lieu of free and open communication, their timidness causes them to bottle up their thoughts and emotions, which then leads to frustration, anger, and passive-aggressive behavior.

In this day and age, it is refreshing to meet an individual who is candid. When they communicate, they possess the four keys of being candid:

> 1. Be direct.
> 2. Be concise.
> 3. Be honest.
> 4. Be genuine.

Successful firefighters are able to express themselves candidly because they know how important it is to communicate effectively and efficiently. In this chapter, we will discuss several qualities of candid firefighters, and why being candid is essential to firefighter success.

BE DIRECT

It's hard to believe, but humans now have an average attention span lasting only 8 seconds, and that is down from 12 seconds from the year 2000.[16]

Life is too short to beat around the bush. When we communicate, let's be direct. When we are, we say what we mean and we mean what we say. We are straightforward with our message, both with our words and our body language. Others know exactly where we stand, because we are not ambiguous or vague. When it specifically comes to requesting what we want, we do not play games. We aren't passive or tentative. Instead, we will tell someone exactly what we want or expect. Our candid communication demonstrates we have confidence and decisiveness.

> "Be direct. Be certain. Be assertive."

Have you ever known a leader who is vague? Someone who is sending a confusing message? How did that make everyone feel? Was it easier or more difficult for everyone to do their jobs?

Leaders must be candid and clear. Their messages, directives, and commands cannot be anything but direct and uncomplicated. Especially when life and death are on the line, fireground leaders cannot leave room for ambiguity.

When delivering expectations and enforcing them, there is no need for leaders to walk on eggshells with their people. When someone under their command is not meeting expectations, a successful leader will be direct with that individual. Instead of sending out a general email or memo to the entire fire department in an attempt to correct a single firefighter's behavior, a true leader will address the deficiency with direct communication. They will sit down with that firefighter and tell them how they are falling short, tell them how to improve, and do it all with respect and guidance. If discipline needs to be administered, the leader will take care of it. In the end, the issue will be corrected because of open and honest communication.

Let's be clear: Being direct does not mean it is acceptable to be rude. We can still be direct while being tactful, respectful, and considerate. Never underestimate the power of the words *"please," "thank you,"* and *"you're welcome."* They really do go a long way when it comes to interpersonal relations.

BE CONCISE

Successful firefighters choose their words wisely. We do not use 100 words to share our thoughts when only 10 words will do. And we do not try to impress others with "$5

---

16 Watson, Leon. "Humans have shorter attention span than goldfish, thanks to smartphones." *The Telegraph*, May 15, 2015, www.telegraph.co.uk/science/2016/03/12/humans-have-shorter-attention-span-than-goldfish-thanks-to-smart/.

words" when simple words are more appropriate. Since we embrace humility, we do not talk "over" people to make them feel unintelligent or small. Whether we are talking with fellow firefighters or citizens, we make sure everyone feels valued and respected with the way we communicate.

While long-winded firefighters love to hear the sound of their own voices, we will communicate efficiently. During training, we do not ask questions we already know the answers to. If we make a comment, it is only to add value to the conversation.

> "Don't be the *askhole*. Don't ask questions you already know the answer to, just to show others how much you know."

If we teach others, we respect everyone's time by delivering the material without unnecessary fluff (e.g. "war stories"). The best instructors are short and succinct, delivering the *what, why,* and *how* that firefighters need to know.

BE HONEST

Honesty can be hard to come by nowadays, but it is critical to candid communication. Honesty is an essential component to worthwhile relationships. It is a building block to so many other core values: *trust, integrity, character, respect, credibility, communication, and many more.*

On professional and personal levels, if we respect someone, we will be honest with them. If they ask for our opinion, we will give them our sincere feedback. We care enough for them to tell them what they *need* to hear, not what they *want* to hear.

> "Honesty is more than not lying. It is truth telling, truth speaking, truth living, and truth loving."
> -James E. Faust

When we hold others accountable, honesty is our only choice. If someone is slacking, we must have the courage to call them out—in a loving and supportive way. Hiding the truth to spare their feelings will do more harm than good. We all know a firefighter who gives an excuse for everything. They have every "reason in the book" of why they are late, why they cannot train, why they cannot work out, or why they forgot to do this or that. If we truly respect them and want the best for them, we will hold them accountable to being better.

As a fire officer, I ask my crew members for their feedback. I ask questions like: *What is going well? What can improve? What would you like to train on?* These questions show that I want their input, and I want to be the best officer I can for them. By soliciting their feedback, I am earning more and more of their trust, respect, and buy-in. Too many officers think they have it all figured out and unfortunately never listen to their firefighters. Eventually, their firefighters will stop listening to them.

However, if we give someone our honest criticism (whether solicited or unsolicited), we must deliver it with tact. We must take their feelings and emotions into account. The goal should be to encourage and improve, not to tear down. Remember the Sandwich Method of delivering criticism. First, praise them for something they are doing right. Then, deliver the criticism tactfully. Finally, praise them again for what they are doing right.

Here is an example of the *Sandwich Method* for a rookie who is failing to fully check their SCBA each shift:

> *"Johnny, I think you are doing a great job by coming in early each morning and putting your gear on the rig right away. When you do, make sure you are doing a full SCBA check to ensure it is all good. And don't forget to log in to it and check the PASS alarm every day. Again, I see you are working hard and fitting in to our crew nicely. Keep it up."*

As we can see, Johnny felt encouraged by the fact we noticed all the good things he has been doing, and now he also knows to do a full SCBA check at the beginning of each shift.

> **"Say what you mean, mean what you say, but don't say it mean."**
> **-Sagar Aswal**

When it comes to disagreements, we must know there is a difference between being honest, as opposed to blaming, shaming, and attacking others. Let's avoid using phrases that include *"you always do that"* or *"you never do that."* Using words like *"always"* and *"never"* to describe others' behavior tend to be overexaggerations that only make the situation worse. Let's remember being candid is not an excuse to be hurtful and offensive when we disagree with others.

Before we even enter a difficult conversation, we must view the situation from multiple perspectives. Let's set our emotions aside (especially anger) and go into the conversation with an open mind, and with the intention to first understand the other person's side of the story.

BE GENUINE

Whether we are giving encouragement, gratitude, apologies, or condolences, we must be genuine. When we are, everything that we say and do is authentic. Being real with others builds trust, confidence, and credibility. However, if we lack sincerity, others will view us as fake and they will see right through our act. If there is no meaning behind our words, those who are listening to us will brush us off.

Being passive aggressive is a form of being disingenuous. When someone is passive aggressive, they lack the courage and character to be sincere with those who are offending them. Such behavior avoids direct confrontation and resolution, and is therefore an immature game that wastes time. Successful firefighters do not play mind games. We do not hint around what we want to change or what needs to change. We communicate exactly what we mean by being genuine and candid. Doing so makes life so much easier for everyone involved.

> **"If you're going to be anything, be genuine."**

When we are genuine, people can tell right away. How we specifically communicate to others matters. Do we look them in the eye? Is our tone of voice positive and affirming? Does our body language show that we care? Remember: *"People don't care how much you know until they know how much you care"* (Theodore Roosevelt). So let's show them that we care.

As we progress throughout our careers, others will appreciate our candor—and we will appreciate others for theirs as well. We will be direct, concise, honest, and genuine, and everyone will know exactly where we stand.

ACTION STEPS

1. Remember the four keys to being candid: *be direct, be concise, be honest, and be genuine.*

2. The next time someone asks for your feedback, be honest and tactful.

3. When you have to deliver bad news, be direct (yet sensitive).

4. Are you sometimes long-winded in your stories and explanations? Practice being concise by getting straight to the point in 30-seconds or less.

SUCCESSFUL FIREFIGHTERS DEMONSTRATE THE BEST IN HUMANITY, ESPECIALLY IN THE WORST SITUATIONS.

FirefighterSuccessBook.com

CHAPTER 16
COMPASSION

HISTORY OF THE MALTESE CROSS

Without a doubt, every firefighter has either worn the Maltese Cross on their uniform or it has been displayed on their fire truck or even their personal vehicle. Long before firefighters started using it, the original Maltese Cross was adopted by the Knights Hospitallers of St. John in the 12th Century.[17] It was later used as a symbol by the Knights of Malta during the Crusades to represent the "Christian Warrior." When fellow soldiers were literally engulfed in naphtha flames during battles, the Knights of Malta would take their capes off and throw them on the flames to extinguish them.[18] With such accounts of the Knight's bravery and compassion, it isn't difficult to understand why the fire service adopted the Maltese Cross.

> "What we do for ourselves dies with us.
> What we do for others lives forever."
> - Albert Pike -
> *(paraphrased)*

---

17 The Maltese Cross: Its Origin and Importance to Malta. MaltaUncovered.com, www.maltauncovered.com/malta-history/maltese-cross/.
18 Kiurski, Tom. "A Piece of Fire Service History: The Maltese cross." *Fire Engineering,* Feb. 9, 2007, www.fireengineering.com/articles/2007/02/a-piece-of-fire-service-history-the-maltese-cross.html.

CHAPTER 16

Figure 16.1 - Original Maltese Cross (Courtesy of Google Images)

Figure 16.2 - The Firefighter Success logo with the firefighter version of the Maltese Cross.

The original Maltese Cross had eight points on it, signifying eight obligations that every Knight of Malta would carry out. Two of these obligations that specifically relate to a firefighter's duties are to *be merciful* and *love justice*—key qualities of compassion.

When any firefighter is asked the reason for becoming a firefighter, 99% of all responses come back to these:

- "I wanted to help others."
- "I wanted to serve my community."
- "I wanted to make a difference in the lives of others."

A successful firefighter is always looking to serve those in need. We sacrifice ourselves so the needs of others come first. Instead of being *selfish*, we are *selfless*—giving our time, talents, training, and even our physical and emotional well-being. Some firefighters have given their lives as the ultimate sacrifice for serving their communities. May they rest in peace, and may we never forget their sacrifice.

COMPASSION FOR THE PUBLIC

To truly achieve firefighter success, we must serve our public with compassion. We demonstrate this compassion by caring for them in their suffering. We give extra consideration to their unfortunate circumstances and do whatever is in our power to alleviate their pain.

> "The purpose of human life is to serve, and to show compassion and the will to help others."
> -Albert Schweitzer

Being compassionate means we are not only sympathetic to those who call for us, but we are also empathetic. We don't just feel sorry for them, but we actually attempt to put ourselves in their shoes and feel what they feel. When we empathize, it opens our hearts and minds to what they are going through. For example, those of us who have experienced a fire in our home are able to easily connect with a citizen who suffers the same. Or perhaps we have personally gone through a cancer diagnosis or have had family members with cancer, and then we respond to a cancer patient who calls for our help. In both scenarios, we empathize with those who are suffering and are better able to serve them.

Even when we don't want to, we are patient with those who aren't necessarily patient with us. We see our citizens when they are in the worst situations of their lives, and they look to us for help. Even if those who call us are drastically different from us, it is still our duty to serve them without judgement or prejudice. Whether they are a different race or religion, gay or straight, rich or poor, etc., we treat everyone with fairness and compassion, because that is what they deserve.

Even if our clientele are "less than honorable" (e.g. dirty, smelly, belligerent, intoxicated, addicted to drugs, physical or sexual abusers, etc.), we will still serve and help them without judgement. I have been yelled and cursed at, threatened, spit on, punched, and more by people on emergency scenes, so I understand how hard it can be to maintain one's poise and still be compassionate.

> "Racism, bigotry, and prejudice have no place in the fire service. It's simple: treat everyone with respect."

As successful firefighters, we will demonstrate the best in humanity, especially in the most inhumane and horrific situations.

COMPASSION FOR OUR BROTHERS AND SISTERS

Compassionate firefighters look out for each other. We pay attention to when a brother or sister is suffering and we ask them to share what is on their mind. We do not wait for someone to ask us for help; rather, we are proactive and aggressive in asking others if we can help. Whether a fellow firefighter is suffering from family or relationship issues, depression, suicidal thoughts, alcohol and drug abuse, etc., we will get them the help they need.

> "Too often we underestimate the power of a touch, a smile, a kind word, a listening ear, an honest compliment, or the smallest act of caring, all of which have the potential to turn a life around."
> -Leo Buscaglia

Above all, we as successful firefighters do not bully our fellow firefighters. Bullying can take up many forms, and some of the offenders will most likely say, *"I was just messing around. It was all in good fun. Everyone was laughing."* But when we witness bullying, it is unmistakable. The victim is embarrassed, quiet, depressed, and usually withdrawn.

Unfortunately, a firefighter brother who I attended the fire academy with was a victim of bullying. After a personal relationship issue and enduring bullying at his fire department, he regrettably took his own life. If we witness bullying at our fire department, we must say something. If we experience bullying from a commanding officer, we must report it to their commanding officer. We must stand up against it with courage. Only cowards bully others to make themselves feel better.

> "Be kind, for everyone you meet is fighting a hard battle."
> - Plato -

COMPASSION FOR OUR FAMILIES

I will be the first to admit that for the longest time, it was difficult for me to come home from a shift at the firehouse and show compassion to my family. Perhaps I ran a lot of calls while I was on-duty or maybe I only got a few hours of sleep. But if I am being honest, the last thing I wanted to do was to come home and be patient with my wife and children.

We must remember that our families need our compassion, patience, and emotional support after being gone for 24 to more than 72 hours. It is extremely difficult for our spouses to do what they do while we are gone. If we have children, we must also remember our spouses are basically a single parent while we are gone. Yes, we have been working hard to provide for our family, but they are the ones who have been taking care of the homefront. Whether they have been taking care of the children, cooking, cleaning, doing laundry, taking care of school functions and sporting events, etc., they have been doing it without our help.

Before we walk in the door, let's take five extra minutes to refocus our minds and attitudes. When we walk in that door, let's be understanding, empathetic, and supportive. When we first see our families, let's tell them that we love them. Let's thank our spouses for taking care of everything. We can seek out ways to serve them by asking our spouses if they need a break from the kids to have some personal time for stress relief.

SHOW EXTRA CONSIDERATION

After a house fire, even if we did an excellent job at the scene, let's give extra consideration to the victims. We must be extra considerate in how we carry ourselves. That is to say, we cannot be "all smiles" and hand out high fives in front of those who have lost their house and possessions. Yes, it is very rewarding for us to use our training and do our jobs well at fires, but we must maintain perspective and balance these emotions in the presence of someone else's tragedy.

GO THE EXTRA MILE

Our words matter. Our actions matter. When others are suffering, the power of a kind word can be incredibly uplifting. When my dad passed away, I still remember those who called me, those who gave their condolences, those who came to his funeral, and those who sent flowers. Their display of compassion meant a lot, and it greatly impacted me during a very difficult time in my life.

Whether it is for our citizens, our fellow firefighters, or our family, let's go the extra mile for them when they are suffering. Let's show our compassion by offering a listening ear, a kind word, or even a hug (when appropriate).

CHAPTER 16

> "No act of kindness, no matter how small,
> is ever wasted."
> - Aesop -

ACTION STEPS

1. The next time you gravitate towards judging a citizen and their circumstances, check yourself. Put yourself in their shoes.

2. Do a random act of kindness for a fellow firefighter, citizen, or a family member. For example, consider leaving them a simple note that acknowledges how hard they have been working.

3. Consider starting a community outreach group at your fire department that aims to help citizens in need.

4. Are you witnessing bullying at your fire department? Stop it in its tracks. Use phrases such as, *"that's enough"* or *"that's inappropriate."* Report bullying to supervising officers.

IT REALLY COMES BACK TO
THE GOLDEN RULE —
TREATING OTHERS HOW WE
WOULD WANT TO BE TREATED.
IF WE CANNOT DO THIS,
THEN IT IS TIME TO SEEK OUT
A DIFFERENT PROFESSION.

FirefighterSuccessBook.com

BE CALM.
BE CONFIDENT.
BE READY.

FirefighterSuccessBook.com

CHAPTER 17
COMPOSURE

VOICES OF EXPERIENCE

John Spera
Firefighter - Paramedic
Co-founder of *Fit to Fight Fire*
Co-author of *Mindset*

The call came in at 12:20 a.m. on July 20, 2012. My crew was dispatched to a single-victim shooting at the Aurora Theater in Colorado. On the way to the call, my mind first went into "skills mode," attempting to remember my trauma training. I remember making it my initial goal to get my patient transported from the scene as fast as possible. Little did I know that I was on my way to America's largest mass shooting at that point in time.

As my engine pulled up to the theater we could see hundreds of people (many covered in blood) who were running toward our truck. A police officer flung open our door and yelled, *"I need your oxygen! I need your oxygen! We have an active shooter inside the theater and there is gas being sprayed."* He was asking for our SCBA and was giving us the first word of an active shooter. My crew got him dressed out and I remember thinking that I was about to be tested.

As I made my way down the east side of the theater, I had both victims and noninjured people running up to me, grabbing me and pleading for help. I was surrounded by hundreds of people—some crying, some screaming, and some helping the wounded.

Many were covered in blood, not from their injuries, but from other victims they had helped. I had to identify who were actually injured, and who were covered in other people's blood. I quickly began to triage and found that the traditional triage system wasn't going to work. Everyone was breathing faster than 30 times a minute and they were all confused.

At that moment, I had to make a choice: *fight, flee, or freeze*. I could let my sympathetic nervous system overtake me, or I could maintain my composure. I could add to the chaos, or I could be the calm. I chose to fight through it.

> "It's our job to respond. It's our job to be the calm during the chaos."
> -Jim Moss

With confidence, I thought to myself: *I got this. I can figure this out.*

When I think back on that call, there were three key elements that provided the foundation to maintaining my composure: *my fitness training, my faith in God, and my family.*

1. FITNESS TRAINING

During my physical training, I would often add fire and EMS scenarios to challenge myself while under physical stress. *Decision making under stress* was something I had read about the military doing and I applied it to our line of work. I used problem-solving during workouts when I was physically elevated and exhausted, and I eventually witnessed it pay off on numerous calls. With my training, I knew I was someone who could work through challenges, adapting and finding solutions as they came about. I believe that continually pushing myself over the years built a mental and physical resiliency that helped me—not only on that call, but also for the years to follow.

2. FAITH IN GOD

My faith gives me meaning to my life beyond my own selfish needs and desires, and a perspective deeper than my experiences. God was with me, helping me to fight for life that morning. He was helping me think clearly, make decisions, and move efficiently. During the chaos, I remember saying this Bible verse to myself: *"I can do all things through Christ who strengthens me."* (Philippians 4:13)

3. FAMILY

Having my priorities in order prior to the Aurora shooting allowed me to be fully present in the situation. I was able to think with a clear head and open heart. Knowing I could be shot and become one of the victims immediately made me think of my wife and kids. Coming home safely to my family motivated me to stay calm and make the right decisions.

We never know when we will face the worst call of our lives, but we have to be mentally and physically prepared for it. Doing so will help us keep our composure so that we can be at our best when others are at their worst.

I am grateful it was me assigned Engine 8 that day, because that is where I was meant to be.

As John shared, when our citizens call us for their emergencies, they expect firefighters who are calm, confident, and even-keeled. Whether it is a routine EMS call or the "biggest fire of our career," we must be level-headed, acting decisively, and moving with purpose. Firefighters who maintain their composure help alleviate their citizens' anxiety and stress—*potentially on the worst day of their lives.*

The public doesn't want firefighters to show up just to *"scream, shout, and throw water all about"* (Chief Rob Fling). They don't want firefighters who are easily rattled or who collapse under pressure. If we lose our cool on the fireground, it communicates to others that we aren't prepared and that we don't know what to do. If our words, tone, and actions demonstrate any lack of confidence, others will lose their confidence in us.

> **"Calm is contagious."**
> **-Rorke T. Denver**
> Commander, US Navy Seal

We are there to mitigate their emergency, not to make it worse. We are there to solve their problems, not add to them.

5 STRATEGIES FOR MAINTAINING COMPOSURE

Here are five practical ways to mitigate the stress that confronts us at emergency scenes.

1. TAKE A DEEP BREATH.

Before we step off the fire truck and go into battle, let's concentrate on what needs to be done. We must calm our minds and prioritize the incident's objectives. If you are the first-due officer who is arriving on the scene of a working fire, take a deep breath before you key up that radio for your initial incident report. When you are ready, talk with clarity and calmness, paint a picture with your words, and confidently give instructions to your crew and incoming units. You will set the tone for the first five minutes, and potentially for the rest of the fire.

2. REMEMBER OUR TRAINING.

Why do elite military forces like the U.S. Navy Seals go into battle with little to no fear? They have spent countless hours training for the mission at-hand. They know the risks and the threats, and they train for them accordingly. They are proactive, not reactive. As Coach John Wooden said, *"When the opportunity arises, it's too late to prepare."* Our mastery of the basics will lead us to success with any emergency that confronts us.

3. REMEMBER OUR EXPERIENCE.

We can use our previous experiences with similar emergencies to guide our course of action. Recognition-primed decision making is what helps us make quick decisions in complex, critical situations.

4. ACT LIKE WE HAVE BEEN THERE BEFORE.

Even if it is our first time forcing a door, stretching an attack line, or throwing a ladder, we will act like we have done it 1,000 times before. We will be confident and execute like the professionals we are. After the fire is out, we won't start handing out high fives and celebrating, even if we did a great job. Let's remember that the victims have just experienced great loss, and we must offer them our compassion.

5. EXPECT OBSTACLES.

How often does everything go "as planned" at emergency incidents? If we are being honest, we can answer that question with: *"almost never."* In order to maintain our composure, we must fully expect the curve balls and then overcome them as they come. A specific tactic that was successful on one fire may not necessarily work on the next fire. In our profession, we must have a Plan A, Plan B, and Plan C. Our ability to adapt will help us maintain our composure and achieve success.

> *"When the opportunity arises, it's too late to prepare."*
> -Coach John Wooden

COMPOSURE IN THE MIDST OF CONFLICT

Maintaining our composure doesn't only apply to emergency responses.

As firefighters who are into the job, we know we have two families: our family at home and our "fire family." Just like our families at home, we will undoubtedly experience interpersonal conflict and disagreements with members of our fire family. Whether it is strategies and tactics, policies and procedures, politics and religion, or even what is going to be for dinner—*we love to debate*.

When we are engaged in a disagreement, we must maintain our composure and be professional. I have witnessed too many firefighters (and even officers) lose their cool and start yelling to try to get their point across. I have even witnessed some firefighters threaten others and become violent. If we have to resort to yelling just to get our point across, others will eventually stop listening.

LEADING WITH POISE

Fire service leaders must be the ultimate example of poise and composure—no exceptions. Whether on the fireground or in the firehouse, firefighters look to them to be their rock. They depend on their leaders to stay calm, to be confident in their decisions, and to lead their firefighters to victory.

Leaders, when those under your command make a mistake, how do you treat them? Do you handle it with poise, giving them the benefit of the doubt? Or do you berate and humiliate them without gaining a full understanding of the situation?

If corrective action is required for their mistake, do you administer the appropriate amount of discipline for the situation at hand? Remember, *discipline* is any action that is used to correct undesirable behavior. Discipline may be something as simple as having a low-key conversation that lasts a couple of minutes. Or the mistake (i.e. violation) may warrant using the fire department's formal disciplinary process.

Whatever the infraction may be, leaders must never yell at their members or embarrass them. Every firefighter and fire officer has made a mistake, and no one wants to be humiliated for their mistake. Let's follow the golden rule and treat others how we would want to be treated—with respect and dignity.

> **"Always maintain your poise. In the firehouse and on the fireground, firefighters want leaders who are calm, confident, and level-headed."**
> - Joe Haefer -
> Battalion Chief, Metro West Fire Protection District (MO)

"KEEP CALM AND CARRY ON."

As we journey throughout our careers, our composure will come from many things: *training, experience, strength, courage, optimism and even our faith.* When others come to us in a panic, we will be the firefighters who are cool, calm, and collected. We will proceed with a quiet confidence that will be contagious to those around us.

NO ONE WANTS A LEADER WHO IS UNSURE OF THEMSELVES. NO ONE WANTS AN OFFICER WHO HAS DIFFICULTY MAKING A DECISION. MOST OF ALL, FIREFIGHTERS DO NOT WANT AN OFFICER WHO SCREAMS INTO THE RADIO AND RUNS AROUND THE FIREGROUND WITHOUT PURPOSE AND DIRECTION.

FirefighterSuccessBook.com

ACTION STEPS

1. Take all training seriously, because we never know when we will need to apply it. Avoid the mindset: *"I don't need to know this. I will never use it."*

2. At emergency scenes, when anxiety levels are high and it is starting to become chaotic, take a step back, think, and ask yourself: *"What is happening and what needs to be done?"*

3. The next time someone yells at you or disrespects you, don't retaliate. Be slow to anger and respond with integrity and a quiet confidence.

THERE ARE TWO THINGS FIREFIGHTERS HATE: CHANGE AND THE WAY THINGS ARE.

- CHIEF ALAN BRUNACINI -

FirefighterSuccessBook.com

CHAPTER 18
CHANGE

For 300 years, the American fire service has been built on tradition—*leather fire helmets, red fire engines, and brass fire poles.* As much as firefighters love tradition, we also secretly want progress and change. It can be very difficult to let go of the traditions that are important to us, but maintaining certain archaic traditions just for the sake of it may not always be the best choice. We must value traditions, and appreciate them for what they are. But we must also make sure that we are not handcuffed by them.

One of the goals of this chapter is to serve as a brief history lesson for firefighters. We will examine how the American fire service has evolved drastically since its inception. We cannot possibly cover everything—our review will be short, yet informative and entertaining.

Most importantly, we will discuss how successful firefighters must adapt to the constantly-changing fire service. Not only must we be adaptable, we are also to be change agents who make a positive impact on our fire departments and communities.

> "But merely being tradition does not make something worthy. ... We can't just assume that because something is *old* it is *right*."
> - Brandon Sanderson -
> *Oathbringer*

CHAPTER 18

A BRIEF HISTORY OF FIREFIGHTING IN AMERICA

The American fire service had its humble beginnings in the early 17th century, when citizens would respond in bucket brigades to wooden chimney fires. In 1678, Thomas Atkins placed the first engine company "in service." Nearly 50 years later, Benjamin Franklin created Philadelphia's Union Fire Company in 1736.[19] Fast forward almost 300 years to present day, it is incredible to see how the fire service has changed—*and continues to evolve*. Let's discuss the numerous ways that the American Fire Service has changed since its inception.

> "The only constant in life is change."
> -Heraclitus

Believe it or not, the fire service did not start in the public sector. It either consisted of volunteers or firefighters who were funded by insurance companies. These companies paid private fire brigades to save as much property as possible. It was not until after the Civil War that government-funded fire departments came about.[20]

The aforementioned "bucket brigades" would consist of two lines of people passing buckets from the water source to the fire. Empty buckets were passed back to be refilled. When the hand pumper came along, buckets were still used to fill it. The foremen of the fire companies would shout orders through "speaking trumpets" to those under their command. That is why modern-day firefighters have bugles as part of their "Class A" collar brass.[21]

Steam fire engines or "steamers" were first pushed and pulled by brute manpower. Fortunately, they were later pulled by horses (which would live in the fire station). It was not until 1910 that gasoline-powered engines became part of the American fire service. The horses, however, were used up until 1923.[22]

Throughout the 20th century, specialized fire apparatus became part of fire departments' resources. These included ladder trucks (aerials, tillers, tower ladders, quints), heavy rescues, water tenders/tankers, wildland and off-road vehicles, small all-terrain vehicles, hazardous materials vehicles, fire and rescue boats, ARFF trucks, rehabilitation trucks, and even air tankers and helicopters.

---

19 Hashagan, Paul. "Firefighting In Colonial America". *Fire Fighter Central*. June 10, 2007, https://web.archive.org/web/20070610030631/http://www.firefightercentral.com/history/firefighing_in_colonial_america.htm. Retrieved Aug., 7, 2019.
20 Smith, Dennis. *Dennis Smith's History of Firefighting in America: 300 Years of Courage*. New York, The Dial Press,1978.
21 Ibid.
22 Ibid.

PERSONAL PROTECTIVE EQUIPMENT

Originally, firefighters had no personal protective equipment, mainly because fires were originally fought from the outside of burning structures. The first uniforms that firefighters actually wore were primarily for parades and ceremonies.[23]

Eventually, firefighters started wearing three-quarters length "hip" boots, long coats (mainly to protect from cold), and the iconic leather fire helmets. These helmets were made with bent rear brims that would keep water and embers from going down their necks and backs. Since then, triple-layer turnout coats and pants have replaced the three-quarters length boots and long coats, providing a drastic change in thermal protection.

James Braidwood invented the first "smoke mask" in 1863, in which fresh air from a bellows was supplied to a firefighter while they were in smoky conditions. Needless to say, these smoke masks did not have longevity in the fire service. It was not until 1945 that self-contained breathing apparatus (SCBA) were introduced by Scott Aviation. The SCBA consisted of a pressurized tank of fresh air (e.g. "room air") that allowed the firefighter to enter a smoke-filled environment.[24] Originally, SCBAs were to be used for "emergencies only," meaning that firefighters could possibly be inside a burning, smoky structure, yet they were not allowed to use them unless *absolutely necessary*. Now firefighters are required to wear their SCBAs in such environments, and even after extinguishment (i.e. during salvage and overhaul).

A firefighter's current PPE ensemble will typically consist of a dozen items or more:

- Helmet
- Hood
- SCBA mask
- SCBA
- Coat
- Pants with suspenders
- Bailout kit with harness, hook, and rope
- Steel-toe boots
- Gloves
- Flashlight
- Pocket tools
- Thermal imaging camera

---

23 "The Evolution of Firefighters' Personal Protective Equipment." *My Firefighter Nation*, April, 26, 2011, www.my.firefighternation.com/group/thegearcrew/forum/topics/the-evolution-of-firefighters#gref.
24 Hasenmeier, Paul "The History of Firefighter Personal Protective Equipment." *Fire Engineering*, June 16, 2008, www.fireengineering.com/index/articles/display/331803/articles/fire-engineering/featured-content/2008/06/the-history-of-firefighter-personal-protective-equipment.html.

A firefighter's full PPE can weigh 70 pounds or more, and a water extinguisher and set of irons (halligan and axe) will add 50 pounds—all of which can take its toll on a firefighter's work capacity and time to fatigue.

RIDING TAILBOARD TO "BUCKLING UP"

Up until the 1980s, firefighters rode the back tailboards of fire trucks during emergency responses. Many firefighters unfortunately died falling off of those tailboards. Eventually all firefighter passengers migrated into fire trucks with open cabs, where they were still exposed to the elements (e.g. cold, rain, snow, etc.). Fire trucks' cabs eventually "closed up," and the next cultural evolution moved toward the use of seatbelts.

Most firefighters tend to believe they are invincible and do not need to wear a seat belt. But as the statistics show, too many firefighters are still getting injured and dying from motor vehicle collisions—both from fire apparatus and private vehicle emergency responses. Within the last few decades, there has been a tremendous push to have firefighters use their seat belts at all times. Many fire departments have delivered special training and required their members sign a seat belt pledge.

COMMUNICATIONS AND TECHNOLOGY

Citizens originally reported fires and requested the fire department by word of mouth. Then the fire alarm telegraph system was instituted in the mid 1800s after the telegraph was invented. These "call boxes" were a more efficient method of summoning the fire department, since an individual would activate it and the signal could go directly to the fire department. This is where the term "box alarm" came from. Firefighters would respond to the box alarm location and get more information on the specific location and nature of the emergency. Due to maintenance concerns, excessive false alarms, and widespread use of the telephone, they lost favor.[25]

With telephones, members of the public were able to directly call a centralized dispatch center, which would then dispatch specific fire department stations and apparatus, depending on the emergency. Telephone lines also allowed commercial fire alarms to be placed inside of buildings. When manually activated, the signal would go directly to the dispatch center, while simultaneously alerting the building's occupants. Now commercial fire alarms are activated automatically by a triggered smoke detector, carbon monoxide detector, or sprinkler system activation.[26]

The way that firefighters receive calls has also changed drastically. When calls would come in, the deafening Klaxon bell would startle a firefighter so bad that their heart would almost come out of their chest. The Klaxon bell has been replaced with pre-alerts and calls that are dispatched over the station PA system.

---

25 "The Fire Alarm Telegraph System." The Firefighter Hall and Musuem www.firehallmuseum.org/about-us-2/about-us/the-fire-alarm-telegraph-system/.
26 Ibid.

Computer-aided dispatch greatly assists dispatchers, who then share call information via mobile data terminals. Now laptop computers and tablets are mounted inside fire apparatus, which provide firefighters with all the information that they need: *scene address, turn-by-turn navigation, nature of the emergency, radio channel, call notes, etc.* Having more information shared via a computer frees up over-the-air radio communications. Instead of saying it over the radio, firefighters press buttons to let dispatch know that they are "on scene," "in service," etc.

Here are other technological advancements that have greatly impacted the fire service, some within the past couple of decades:

FIRE SPRINKLER SYSTEMS

First installed in the early 1800s, fire sprinkler systems have dramatically improved both occupant and firefighter safety. Getting water on the fire as soon as possible, automated systems became popular for all types of occupancies (not just commercial) in the mid 1900s. Now it is standard building code to have them present in public buildings like schools, hospitals, churches, and hotels.[27]

THERMAL IMAGING CAMERAS

TICs allow us to see heat in ways the naked-eye cannot. Whether we use them for size-ups, overhaul, search and rescue, finding energized power lines, locating an overheated piece of electrical equipment (and more), the thermal imaging camera is a tool that has improved the way we do the job.

TRAINING DELIVERY

For the longest time, firefighter training was primarily delivered on the training ground and fire station bay floor. With the arrival of the internet, online and computer-based training has become more popular. We are now in the age where virtual reality training is an option. However, we must remember that computer-based training is an additional tool and can never replace hands-on training.

DRONES

Also known as "unmanned aerial vehicles," drones have been used by the military for quite some time. The fire service is now using them for size-ups and reconnaissance at structure fires and wildland fire operations. An incident commander can use them to monitor the progress of firefighters performing roof-top/ventilation operations. Drones with thermal imaging cameras are even used for large area search operations, especially at night.

---

[27] Bellis, Mary. "A Brief History of Fire Sprinklers." *ThoughtCo*, Feb. 11, 2019, www.thoughtco.com/fire-sprinkler-systems-4072210

FIRE STATION DESIGN

Horses once lived in the fire station with firefighters. A dalmation was the iconic firehouse pet. The kitchens and lounges were primarily in the apparatus bays, exposed to harmful diesel exhaust. Firefighter sleeping quarters consisted of a large open bunk room without dividers. If you weren't running calls at night, you were kept awake by one of your crew member's snoring.

A lot has changed in the last century with the way that fire stations are designed. Large bunk rooms have slowly become individual rooms. Firefighter PPE that was typically housed in the apparatus bay (exposed to diesel exhaust) is now migrating into a separate, negative-pressure room. Diesel-exhaust capturing hoses connect to the fire trucks and remove carcinogenic toxins from the apparatus bay. It is now becoming the norm to have commercial extraction washers for contaminated PPE *and* residential-style washers and dryers for station uniforms. Firehouse kitchens and lounges have moved out of the bays, allowing firefighters to eat and rest without worrying about carcinogenic diesel fumes.

WOMEN AND MINORITIES IN THE FIRE SERVICE

For the longest time, the fire service only consisted of white men. Unfortunately, the mindset of *"women have no place in the fire service"* was common. As the civil rights movement paved the way for the inclusion of more minorities, the American fire service has slowly become more diverse over the past half century.

Within the past few decades, fire departments have come to the realization that the make-up of their firefighters must reflect the diversity of the public that they serve. That is to say, fire departments must become more diverse and inclusive of women and all races. Aggressive fire departments have hosted specific outreach events to recruit more women and minorities. Progress has come slowly, but the American fire service now consists of 7% women, and 22% non-whites[28].

SMOKING AND ALCOHOL IN THE FIRE STATION

Isn't it ironic that the same firefighters who battled fire, heat, and smoke used to take a cigarette break after the fire was out? And then they would come back to the station and continue to smoke inside the firehouse? I have heard plenty of stories of how the cigarette smoke was so thick in the station that firefighters had a hard time seeing each other. That may be an exaggeration, but it is true that smoking cigarettes, pipes, and cigars was common in the station's living area and apparatus bay. As the American culture changed its view on tobacco, smoking was prohibited inside the station. Now some fire departments go all the way to prohibit their personnel from using any tobacco products, *both on and off duty*.

---

28 Evarts, Ben and Stein, Gary P. U.S. Fire Department Profile 2017. National Fire Protection Association, 2019.

It may seem absurd, but firefighters drinking alcohol in the fire station used to be a regular occurrence. That's right—the very driver whose responsibility it was to get the engine and crew to the scene safely could have been under the influence of alcohol. The same could have been said for the firefighters who were responsible for saving lives and property. Needless to say, fire departments now prohibit their personnel from using alcohol while on duty, and put restrictions on volunteer firefighters responding to emergencies if they have alcohol in their system. Fire departments now use randomized drug and alcohol screening, as well as screening after a fire department apparatus has been involved in a collision.

CANCER PREVENTION

"Salty" gear was once a badge of honor for firefighters to show that they were fighting fire. Firefighters believed that *"dirtier was better."* If a firefighter's gear (and subsequently the station) smelled like smoke, it was a source of pride. In the past, firefighters would keep their contaminated gear in the bunk rooms (e.g. "bunker gear"). Showering after a fire was uncommon, because no one wanted to lose that post-fire cologne.

The past couple of decades have shown us contaminated PPE and contaminated firefighters only lead to cancer. Depending on the type of cancer, firefighters have twice the risk (or more) of contracting it, as compared to the general population. The Boston Fire Department has shared that a Boston firefighter is diagnosed with cancer every 3 weeks[29]. Due to the modern fire environment, petroleum-based materials (i.e. plastics) and flame retardants are overly abundant, and the products of combustion are a toxic nightmare.

There have been great strides in cancer prevention, and here are some of the steps firefighters are now taking to reduce their risk:

- Breathing from their SCBA during suppression and overhaul efforts
- Gross decon of their gear with water at the scene
- Wiping their skin with wipes after a smoke exposure
- Showering after a smoke exposure
- Laundering their contaminated uniforms
- Laundering their contaminated PPE in an extraction washer (i.e. "gear washer")
- Fire departments are purchasing a second set of turnout gear for their firefighters
- Regular exercise, improved nutrition, and controlling their weight
- Annual NFPA 1582-compliant medical evaluations with cancer screenings
- Diesel exhaust systems in the apparatus bay
- PPE stored in a room away from the apparatus bay (and not in sleeping quarters)
- Regularly decontaminating tools, equipment and the fire truck's cab

---

[29] Roman, Jesse. "Facing Cancer." *NFPA Journal*, May 1, 2017, https://www.nfpa.org/News-and-Research/Publications-and-media/NFPA-Journal/2017/May-June-2017/Features/Facing-Cancer.

Since cancer is one of the leading killers of firefighters, it is no surprise that so many firefighters have become extremely aggressive with decontaminating themselves after a smoke exposure.

POST-TRAUMATIC STRESS, SUBSTANCE ABUSE, AND SUICIDE

As firefighters, we respond to and see some absolutely horrific scenes. With these events, some firefighters are able to process them in a manner that has a minimal effect on them. Yet for others, the same type of call can leave a lasting, devastating impact. We now know that the traumatic events we experience in our profession leave mental and emotional wounds that may eventually turn into long-term scars.

Post-traumatic stress (PTS) and mental illness in the fire service are real. Every day, more firefighters are being diagnosed. Year after year, the number of firefighter suicides outnumbers the amount of line-of-duty deaths. Firefighters have a higher prevalence of PTS, depression, alcohol abuse, suicidal ideation, and suicide than the general population.[30]

> "I wish my head could forget what my eyes have seen."
> -Dave Parnell
> Detroit Fire Department

Fortunately, more firefighters are getting the help that they need. It was once taboo for a firefighter to admit that they were suffering. But now the fire service has peer support teams, specially trained counselors for firefighters, and even specialty treatment centers.

If you or someone you know is suffering from any of the aforementioned issues, refer back to *Chapter 6 - Courageous* for a list of helpful resources.

EMERGENCY MEDICAL SERVICES

EMS had its beginnings after combat medics came back from the Vietnam War. Although very rudimentary and simple, the first EMTs would simply place their patients in a makeshift vehicle (sometimes a glorified station wagon) and bring them to the hospital. Over the past several decades, prehospital patient care has become extremely advanced, with paramedics performing intubations, administering a full gamut of medications, and even performing CT scans in the field.

Much like the fire service's initial view on women and minorities, EMS was seen as having "no place" alongside firefighting.

---

[30] Heyman, Miriam, et. al. "The Ruderman White Paper on Mental Health and Suicide of First Responders." Ruderman Family Foundation, April 2018.

But now, many fire departments have firefighters who are also certified as EMTs, paramedics, and even registered nurses. Depending on the fire department, medical calls make up 70—80% of all emergency responses. Fire departments even collect revenue from patient treatment and transport, which helps support their budget.

With all of the advances that have been introduced over the past century, it is intriguing to see how the fire service will evolve over the next hundred years.

CHANGE OR DIE

The quote at the beginning of the chapter hits the nail on the head. Most firefighters are not satisfied with the status quo, and yet they have a difficult time navigating the ever-changing fire service. However, if we are to be successful, we must embrace change openly and quickly. As we discussed in *Chapter 1: Coachable*, successful firefighters are those who are *teachable*. If we are teachable, it demonstrates that we are open-minded. If we are open-minded, then we are able to understand the "why" behind the changes that we are confronted with. If we are able to understand, then we can finally accept change and readily adapt.

During our time in the fire service, change will occur both individually and organizationally. As individuals, we will get older, slower, and more prone to injuries. But on a positive note, we will also become wiser, more experienced, and better trained—which we will use to our advantage.

> "We cannot become what we want to be by remaining what we are."
> -Max Depree

Outside of the fire department, we may experience a great deal of change: getting married, having kids, moving homes, illness, divorce, etc. All of these changes and how we respond to them will affect our roles as firefighters. If changes at home become too much for us to deal with, we will not be able to be successful as firefighters. That is why it is critically important to have a strong support network of family and friends (outside and inside of the fire department) to help us when times get hard.

> "Some work harder, but successful firefighters work smarter."

Organizationally, change can come from the administration level, and it can also start at the grassroots level. It can come from outside of the fire department (e.g. NFPA standards, state mandates, lack of support from the community, etc.), or it can start within our own fire department and spread to other departments around us. Whatever change comes our way, our success will depend on our ability to adapt.

WHAT WILL HAPPEN IF WE ARE RESISTANT TO CHANGE?

As Charles Darwin said, *"It is not the strongest of the species, nor the most intelligent that survives, but rather the one that is most adaptable to change."* If we cannot adapt to change in the fire service, there are two possible outcomes: We will stay in the same place, doing the same things; or our lack of skills and knowledge will actually cause us to regress.

By not adopting new technologies and best practices, we will be providing our citizens with a substandard level of service. Or perhaps even worse: we could be jeopardizing their safety and our safety.

WHY ARE PEOPLE RESISTANT TO CHANGE?

Fire Chief Marc Revere shared that individuals may present three types of resistance: *Not knowing, not able,* and *not willing.* With the first category, such people do not understand why the change is taking place. Perhaps "the why" was not properly explained to them, or perhaps they do not have the mental capacity to understand the change.

With the second category, "not able," these people are so entrenched in their ways they have a mental roadblock that prevents them from getting on board with change. This could be the firefighter who has been on the job for decades, doing the same exact things the same way. For them, a big change may prove to be too difficult.

Finally in the last category, "not willing," this is the person who simply refuses to adapt to change. Their unwillingness to comply comes from the mindset we have all heard before: *"But we have always done it this way. I'm not going to change now."* Unfortunately, this last group may only have three options for dealing with organizational change: retire (if they are able), quit, or face disciplinary action, which may involve termination.

BE THE CHANGE

If we are to truly achieve success, it must go further than just our ability to welcome change. We must be change agents who inspire and lead positive change in our fire departments and communities. As catalysts for progress, we examine our culture to determine what needs improvement. If we see a problem, we do not merely complain about it—we come up with a solution and we carry it out. We are the change we want to see in our fire departments.

One of the primary reasons I wanted to become a company officer was to have a better platform to inspire positive change. After being promoted to the position of lieutenant, I wrote several fire department policies (i.e. cancer prevention), took organizational health and fitness to the next level, and improved EMS training delivery. As a captain, I am always looking to improve my crew's knowledge, skills, and abilities through training, professional development, and personal development.

FIREFIGHTERS WHO ARE NOT ABLE TO CHANGE AND THOSE WHO ARE NOT WILLING TO CHANGE WILL NOT SURVIVE THE FIRE SERVICE.

FirefighterSuccessBook.com

Every single firefighter can be a change agent. Regardless of rank or time on the job, If we are passionate about inspiring positive change, we will make it happen. It may not always happen as fast or as big as we want it to, but if we are determined and persistent, we will make waves.

> "Be the change you wish to see in your fire department."
> -Mahatma Gandhi
> *(paraphrased)*

8 KEYS TO LEADING CHANGE

1. START SMALL.

Rome was not built in a day, and we will not accomplish significant, lasting change in just a day. Depending on the type of change we want to inspire, we must take small, deliberate steps towards our goal. We must develop a plan and see the long-term goal. Along the way, we can achieve smaller milestones which will eventually culminate in our plan's success.

2. COMMUNICATE AND INFORM.

As we have discussed, people need to know (and deserve to know) *why* a change is necessary. They also want to know how it will directly impact them. As agents for change, we must prove the change will improve their lives or make operations more effective and efficient. We will be clear, concise, and to the point.

3. LEAD BY EXAMPLE.

If we are going to *talk the talk*, we must *walk the walk*—always. If we are going to preach that we want positive change, we have to live out our message 100% of the time. If we do not, we will lose credibility and our plan will lose credibility. Personally, I am someone who promotes fitness within my fire department (so much so that I wrote a book on it). But, if I demonstrate to others that I do not exercise, eat right, and gain 100 pounds, then my message has no credibility and I will not inspire anyone to change.

4. CONSULT WITH KEY PLAYERS AND WIN THEM OVER.

Some may call this "playing politics," but in order for our plan to be successful, we must consult with the key decision makers and convince them. If it is a change that doesn't directly impact the chief, yet improves the working conditions and morale of the operational staff, we must prove how it will benefit the entire fire department and the community.

5. CHIEF FRANK VISCUSO: "THE BEST IDEAS HAVE TO WIN."

If we truly want the best for our firefighters and the organization as a whole, the best ideas have to win. Perhaps our plan has a few shortcomings, and someone else came along and improved it with their ideas. Even if they get most or all of the credit, we must be humble and put our pride aside. Let's have the humility to accept the fact that their idea was better. Let's compromise for the greater good.

6. COMPANY OFFICERS ARE CRITICAL.

If change originates at the administrative level and needs to be delivered to the "rank and file," company officers are the most powerful tools to achieve it. Company officers must have 100% buy-in to deliver change, or else their members will not get on board.

7. DO NOT UNDERESTIMATE YOUR LEVEL OF INFLUENCE.

Whatever influence you have, use it. You may not have rank, title, seniority, or much time on the job. Don't worry—you can still drive positive change at your current level. It may take a grassroots effort to convince others, but it will be worth it.

8. BE PATIENT.

Change takes time, energy, and passion. Don't stop. Great firefighters and fire departments aren't cultivated in a day. It may take months or years (if not decades) to achieve noteworthy improvements. If the change we seek is worthwhile, we must be persistent, yet patient.

A WORD OF ENCOURAGEMENT

If we are striving for positive change in our fire departments, be forewarned: Since we are challenging the status quo, we will undoubtedly face adversity from those who like things just the way they are. These firefighters will gossip about us and attempt to slander us because they view change as a threat. Pay them no mind, and stay the course.

In the fire service, change is inevitable. Since its beginnings 300 years ago, firefighting in the United States has continuously evolved and it will continue to do so. Successful firefighters not only embrace and adapt to these changes, they also lead positive change for their fire departments and communities.

> "Progress is impossible without change."
> - George Bernard Shaw -

CHAPTER 18
ACTION STEPS

1. Research the history of your own fire department. When was it founded? What were some important milestones? How has it changed?

2. What is something that needs to change at your fire department? What can you realistically do to change it? How will you start?

3. Focus on what you can actually control. Inspire small unit change within your immediate circle of influence, and then watch it eventually spread to other people outside of your circle.

4. Refer back to *Chapter 7 - Conviction*. Based on your responses to the questionnaire and the *Personal Strategic Plan* you created, what is one thing that you want to change or improve about yourself? Set a SMART goal to achieve it.

AS CHANGE AGENTS FOR THE FIRE SERVICE, WE ARE NOT SATISFIED WITH THE STATUS QUO OF MEDIOCRITY. WE ARE NOT SATISFIED WITH ONLY ACHIEVING THE MINIMUM STANDARD. WE STRIVE FOR CONSTANT IMPROVEMENT, INDIVIDUALLY AND FOR OUR FIRE DEPARTMENT. WE STILL VALUE TRADITION, BUT WE ARE ALWAYS ASKING: HOW CAN WE IMPROVE? HOW CAN WE BE BETTER, FASTER, STRONGER, SMARTER, AND SAFER?

FirefighterSuccessBook.com

FIREFIGHTING TAKES A TEAM. WE CANNOT ACHIEVE SUCCESS ON OUR OWN.

FirefighterSuccessBook.com

CHAPTER 19
COMMUNITY

No matter how hard we try, we cannot truly achieve success if we try to do it all on our own. As the African proverb tells us, *"If you want to go fast, go alone. If you want to go far, go together."* It is only through a strong sense of community that firefighters, fire crews, and the fire department can work together as a team to fulfill their mission of serving the public with excellence. In this chapter, we will discuss how successful firefighters build successful teams through three essential components of community: common purpose, camaraderie, and company pride.

COMMON PURPOSE: THE MISSION COMES FIRST

The mission must always be the team's most important focus. Whether it is training, responding to calls, public relations events, or daily operations like truck checks and cleaning—the mission is critical, and it is what fuels us. The mission is what gives us a common purpose.

> "The most important thing is for a team to come together over a compelling vision, a comprehensive strategy for achieving that vision, and then a relentless implementation plan."
> - Alan Mulally -

Ask yourself: *Why am I a firefighter? Why did I take the oath? What is my mission? What is my purpose?* If your answers fall back to recognition or money, it is time to re-evaluate your mindset and motivation, or perhaps it is time to find a different vocation altogether.

> "Every firefighter's mission is simple: *Others are first, I am last.*"

Everything we say, do, and deliver must come back to the ultimate goal of serving others with passion, excellence, and sacrifice.

Are we responding to a structure fire with victims trapped? *We serve them.*

Are we training on fire, rescue, EMS or fitness? *We serve them.*

Are we performing a blood pressure check for a station walk-in? *We serve them.*

Are we out delivering public education or community relations? *We serve them.*

Whatever we do, we do it to serve, protect, and care for others. *It is all for them.*

GETTING EVERYONE ON BOARD

As a team of firefighters, understanding the mission and buying in is critical to building community and teamwork. Sharing a common mission means we share a common purpose. When we are all on the same page, the team's success is inevitable.

If we are leaders in our fire departments (formal or informal), it is critical that we communicate clear expectations to our people. It is important everyone knows the fire department's (or at least leadership's) core values, mission, vision, and goals. In our profession, it is imperative that all leaders communicate and expect the following SECRIT core values from our teams:

- **Service:** We put others and their needs before our own.
- **Excellence:** We display the highest level of professionalism.
- **Compassion:** We care about our citizens and each other.
- **Respect:** We treat people right.
- **Integrity:** We always do the right thing.
- **Trust:** We earn trust through our actions, words, and behaviors.

As leaders, we must demonstrate these values every single day. If not, those under our influence will never buy-in to the mission. By living them out, we are displaying what we want others to emulate. Those under our leadership are always watching; we cannot underestimate our level of influence. Our actions are contagious.

If we have team members who are having trouble getting on board with the mission, let's find out why. Let's ask them questions and get them to talk. Perhaps they had been "burned" in the past by a bad experience with a company officer or chief. Perhaps they have had poor leadership for such a long time they have become apathetic or jaded. Let's show them what it means to have passion for the job again. Let's treat them with the SECRIT core values and watch how they change.

There isn't one firefighter who started their career with a poor attitude towards the job. Every single one of us started with excitement and passion. No one is a lost cause, and it only takes the right leader to bring their fire back. Be that leader.

> "The fire service didn't issue you a draft card. You chose to be here. Act like it."
>
> -John Dixon
> **Battalion Chief, Teaneck Fire Department (NJ)**

CAMARADERIE

To build a true community, we must invest in each other by legitimately caring about each other. When we do, it makes an undeniable difference in building trust, respect and relationships—*the non-negotiable building blocks of high-performance teams.*

Here are 12 tried-and-true ways to grow a team through camaraderie:

1. GET TO KNOW EACH OTHER.

Find out each other's likes and dislikes, their strengths and weaknesses, accomplishments and failures, etc. If we don't truly know each other, how can we build trust, relationships, and community?

2. SPEND TIME TOGETHER WHEN ON DUTY.

Eat together, train together, work out together, etc. All of these go a long way to building teamwork and chemistry. If we don't spend any time "doing life" together, it will be extremely difficult to come together as a team.

3. SPEND TIME TOGETHER WHEN OFF DUTY.

Grab some coffee, share a meal or a drink. Occasionally plan a fun activity: play or watch sports, attend a concert, etc. Social events outside of work tend to break down the invisible barriers that are typically present in the workplace. Consider attending an off-duty training opportunity together.

> "We are a team—we train together, we eat together, we laugh together, we struggle together."

4. STAY CONNECTED.

Be on a text message group together to stay connected when everyone is off duty. Be forewarned: It may become hilarious and get a little out of hand.

5. MAKE IT A POINT TO KNOW YOUR CREW'S FAMILY MEMBERS' NAMES.

As silly as it may sound, put their family members' names in your phone when you learn them. Later on, ask about each others' family, using specific names. *Was someone's son recently sick? Did someone's daughter just graduate high school?* Our team will notice when we ask about their family members.

6. RECOGNIZE SUCCESSES AND MILESTONES.

A true teammate will celebrate other team members' successes. Did a fellow firefighter get a promotion, award, or accolade? Let's be happy and congratulate them—even if they were our "competition" during the promotional process. Along the same lines, let's know our crew members' birthdays and work anniversary dates so we can recognize them. A simple gesture like getting a dessert to celebrate will go a long way towards building team camaraderie and morale.

> "If everyone is moving forward together, then success takes care of itself."
> -Henry Ford

7. HOLD EACH OTHER ACCOUNTABLE.

If someone is falling short or not doing their fair share, have that tough conversation with them. Our words may be hard to swallow at first, but if we approach the situation with respect, trust and honesty, the other person will appreciate us for it. *"As iron sharpens iron, so one person sharpens another." (Proverbs 27:17)*

8. HAVE EACH OTHERS' BACKS.

Is someone bad-mouthing or falsely accusing one of our crew members? We will stick up for our team. We won't let others gossip or spread rumors about them.

9. FIND SPECIFIC WAYS TO HELP EACH OTHER OUT, BOTH ON AND OFF DUTY.

As Jim Stovall has shared, *"When we all help one another, everybody wins."* One of my crewmates and his wife tragically experienced a miscarriage. I made sure to frequently reach out to him, and I cooked a dinner and brought it to them so they would not have to worry about cooking for one night. It was no amazing gesture, but it was something small and tangible to show that I cared. They greatly appreciated the meal and it showed that I was thinking about what they were going through.

10. ASK FOR HELP.

Sometimes all we need to do is just ask. Do we need help with a project at the firehouse, or perhaps a house project? Do we need help moving into a new house? Or maybe we need emotional support because of a traumatic call or a serious issue at home. When it comes to the latter, this will require us to be vulnerable because we are saying that we cannot do it on our own. 99% of the time, our team will rally around us with full support when we ask for help.

11. HAVE AN ATTITUDE OF GRATITUDE.

A simple *"thank you"* goes a long way for someone who cooked dinner, led training, or folded our laundry. Even if it is the rookie's responsibility to make coffee and clean the dishes, we can still thank them. Everyone wants to feel appreciated and respected for their efforts.

12. DISAGREE WELL.

Not everyone will agree on everything. Learn how to respectfully disagree, and when necessary, compromise for the greater good of the group. Remember: Sometimes it is better to be kind than to be right. Let's not let a simple disagreement or a minor misunderstanding turn into something that divides the team.

> "It is amazing what you can accomplish if you do not care who gets the credit."
> - President Harry S. Truman -

COMPANY PRIDE

Nothing builds community amongst firefighters more than company pride. As firefighters, it goes without saying that we are a competitive bunch. We expect our company and station to be the best. We seek to prove our skills and knowledge on the fireground every opportunity we get. We want to be known as the go-to crew that gets the job done better and faster than everyone else. Some may call this ego, but we know it to be company pride.

CHAPTER 19

Here are four ways to build community through company pride:

1. Keep a clean firehouse, truck, and tools. Consider painting your hand tools with accent colors or numbers that are unique to your company.
2. Check your equipment every shift.
3. Do hands-on training together every shift.
4. Do a special project together. Consider creating a company or station logo, a firehouse t-shirt that is unique to your crew, fire station wall art, or a custom firehouse kitchen table.

A thin red line firehouse kitchen table, created by Capt. Jim Moss and his crew members.

COMMUNITY

*An American flag made of red, white and blue fire hose.
Created by Assistant Chief Brian Zaitz and members of the Kirkwood Fire Department (MO).*

A custom logo from Metro West Fire Protection District Station 5 (MO).

CHAPTER 19

FIREFIGHTER SUCCESS: IT TAKES A TEAM

This book is primarily centered on you, the individual, achieving firefighter success at a personal and professional level. However, it is ironic that in order for us to achieve such success, we must be part of a community and invested in the team. Napoleon Hill tells us: *"It is literally true that you can succeed best and quickest by helping others to succeed."* If we want to be successful in the long-term, we will share a common purpose: the mission to serve others. We will build a great team and achieve the mission by investing in each other.

> "Individual commitment to a group effort—
> that is what makes a team work."
> - Vince Lombardi -

ACTION STEPS

1. Cook a meal for your crew at the firehouse.

2. Plan and lead a workout and a training for your crew to do.

3. Plan an off-duty outing for your crew.

4. Learn the names of your crew members' family.

5. Say *"thank you"* as much as possible.

6. Complete a "company pride" project listed above.

THERE IS NO SUCH THING AS A SELF-MADE FIREFIGHTER. YOU MAY BE EXTREMELY INTELLIGENT, YOU MAY BE TALENTED, YOU MAY BE SKILLED — BUT IF YOU ARE GOING AT IT ALONE, YOU WILL NEVER FULLY ACHIEVE FIREFIGHTER SUCCESS.

FirefighterSuccessBook.com

IF WE WANT TO SUCCEED, WE WILL HELP EACH OTHER SUCCEED.

FirefighterSuccessBook.com

CHAPTER 20
COACH

Everything comes full circle in this, our concluding chapter. Throughout this book, we have discussed the tools and mindset that will help every firefighter achieve success, and now we will focus on making sure that we equip other firefighters on *their* journey to success.

If we are honest, we will all admit we need others' help in order to be successful. And for those of us who have been in the fire service for a little while, we know helping others achieve their success is a crucial element to achieving our own success. Let's examine what a coach does and how we can become the best possible coaches for firefighters seeking to succeed.

> "The greatest success we'll know is helping others succeed and grow."
> -Gregory Scott Reid

WHAT DOES A COACH DO?

As the years go by and we continue to invest ourselves in the fire service, we come to realize that we must own a great responsibility: *to train, teach, and mentor the next generation of successful firefighters*. In simple terms, we are to be coaches to new and progressive firefighters, so that they keep the fire service an honorable calling. Through the experience and wisdom we have gained throughout the years, we will show the way so that every firefighter under our influence can *go the way*. Undoubtedly, others have done this for us, and we will pay it forward for the next generation.

> "A star wants to see themselves rise to the top. A coach wants to see those around them rise to the top."
> -Simon Sinek
> *(paraphrased)*

To be the best coaches possible, we must have the ability to see the bigger picture. In doing so, we are able to see the greater potential in our firefighters, even beyond what they see and believe about themselves. Not only do we see their potential, we help them achieve it through training, teaching, and mentoring. We know our firefighters' strengths and weaknesses, and therefore we know when to push them, when to encourage them, when to hold our them accountable, and when to expect more from them.

A good coach's forward-thinking mindset allows them to visualize successful outcomes long before they happen. We know what we want the end result to be, so we plan accordingly and take the appropriate steps. We are also able to balance long-term goals with short-term objectives. We do not get lost in minute details that could possibly distract us from what is actually important. When it comes to our team, we delegate the right tasks to the right people because we know their talents, knowledge, skills, and abilities.

Balancing both optimistic and realistic perspectives, we are able to adequately guide our members in their career choices. Let's be honest, there are some firefighters who want to become officers, but it is obvious that they do not possess the leadership and managerial qualities needed for the position. A good coach will help them develop and maximize the strengths they do have and guide them appropriately to make the best possible career choices. Perhaps such members are better suited as senior firefighters or driver/operators.

It is perfectly acceptable to say that not everyone should become an officer, a chief, a driver/operator, or a _____ (fill in the blank). Similar to a football team, not everyone is suited to be a quarterback, a lineman, a safety, a kicker, etc. However, the best coaches help their people maximize their potential and be successful.

> "Doing the best we can with what we are given— this is the definition of success."

MENTORSHIP – A PERSONAL STORY

I have been incredibly fortunate in my career. I have had not just one, but several mentors who have greatly invested in me during my journey to success. They have poured their knowledge, experience, and wisdom into me. Most importantly, they gave me their *time*—something that is absolutely priceless in this day and age. To help other firefighters succeed, we must devote our time to developing them and investing in them.

When I was an aspiring officer, my mentors taught me how to complete the managerial tasks of an officer (e.g. prioritization of duties, organization, entering manpower in the daily schedule, completing training records, etc.). They also trusted me to "ride the seat" on the fire truck as the acting officer, while they rode as the backstep firefighter. They taught me things that were never discussed in my fire officer certification classes and books. They had high expectations for me, and they pushed me to be the best I could be.

Each of my mentors wanted the best for me, and they genuinely wanted me to succeed. I am forever grateful for their leadership, mentorship, and time; without them, I would not be where I am today.

As coaches and mentors, we must have the same vision and drive for our firefighters.

10 STEPS TO BECOMING AN EFFECTIVE COACH

1. SHOW THAT WE CARE.

We must care about our firefighters and know *who* they are before we focus on *what* they can do (i.e. knowledge and skills). As coaches, let's first build a foundation with relationships that are based on trust and respect. After we figure out our members' strengths, weaknesses, goals, etc., we can then guide, encourage, and challenge them along their journey.

2. OBSERVE AND LISTEN.

One of a coach's greatest tools is the act of observation. The second greatest tool is listening. Not surprisingly, these two go hand-in-hand. When we employ them, we learn so much about our team members. When our people have a problem, they don't always need our advice. Sometimes all they need to do is talk about it and vent. Often, they will come up with their own solution, without much (or any) of our input.

> "When we coach, we aren't raising others up to our level. We are expanding their potential and helping them achieve an even higher level of success than we ever could."

3. BE A VISIONARY.

We will visualize the end result and the success that our team will achieve. Once we foresee the end result, we can develop a step-by-step gameplan. Let's help our members set goals and push them.

4. KEEP COACHING SIMPLE.

In every area of our profession, we must teach the *"what, why, and how."* That is to say, we first teach our firefighters the basics of *what* they need to know to succeed. Then we teach them *why* it is so important that they know it. With the "what" and "why" in place, we can then instruct them on *how* to accomplish the task at hand. When it comes to the "how," sometimes it is actually best for them to do it the way they see fit, depending on their knowledge, experience and skill set. They may make mistakes, but their mistakes will be very valuable lessons in their development and journey.

5. INSPIRE PASSION.

Brian Cagneey said it best: *"Coaches are aware of how to ignite passion and motivate people. They have an energy that is contagious and know exactly how to get their team excited."* To inspire passion in others, we must first demonstrate our passion for the job. Let's ask ourselves: *Are we just on the job, or are we into the job?* Every day we must show up with a positive attitude, encouraging and expecting our firefighters to grow. Next, we must get our members to buy-in by figuring out what motivates them. Not everyone is motivated by the same methods; that is why it is so important for us to really know our people. The 20-year firefighter will be inspired differently than the rookie, so we will have to do some investigating, as well as trial and error.

6. BE A MENTOR.

In the fire service, mentorship relationships are critical to succession planning. We must always have the mindset of *"training our replacement"* (Chief Tiger Schmittendorf). Such coaching relationships are special because they connect the mentee and the mentor in a one-on-one relationship. As mentors, we will teach them countless lessons that cannot be learned in books or in the fire academy. We will use our personal training, education, and especially our experience to train them up to be the best versions of themselves.

7. FOLLOW UP REGULARLY.

Let's check in with our people often. If they are taking training classes or working towards specific firefighter certifications or a college degree, ask them how everything is going. Ask them how many classes they have left before they graduate. See if they need any help. Maybe they are slacking a little or feeling unmotivated, and our questions will prompt them to get back on the horse and kick it in high gear.

8. HAVE THE TOUGH CONVERSATIONS.

There will be times when we need to tell our members things they don't want to hear. Whether they are underperforming intentionally or unintentionally, we must have the courage to hold them accountable and then redirect them on how to improve. Let's remember to focus on correcting the undesirable behavior, and not to tear down the person themself.

9. HAVE THEM LEAD.

Nothing will have our people progress faster than having them lead. Do they want to become an officer? Have them regularly "ride the seat" as the acting officer. Do they want to become a driver/operator? Have them drive the fire truck often and be the "acting driver/operator." Have them lead company training on a wide variety of topics. Have them lead public education and public relations events. All of these opportunities will help them to grow their confidence and abilities exponentially.

10. RECOGNIZE THEIR SUCCESS.

Whether it is a small or large achievement, it is crucial that we celebrate everyone's success. Use simple statements like:

- "I saw how hard you worked on that. Keep it up."
- "I'm proud of you."
- "I had no doubt you could do it."
- "Great job with _____."

10 QUALITIES OF UNSUCCESSFUL COACHES

Now that we have discussed 10 ways we can become better coaches, let's quickly examine 10 qualities of unsuccessful coaches.

Unsuccessful coaches do the following:

1. Care more about themselves than their team.
2. Don't accept the team's input.
3. Coach with fear and intimidation.
4. Don't lead by example.
5. Don't deal with conflict.
6. Don't delegate tasks and never ask for help.
7. Don't share their knowledge, for fear that their members will become smarter than them.
8. Are overbearing, controlling, and micromanaging.
9. Take all the credit for success and blame everyone else for failures.
10. Are pessimistic, negative, and apathetic.

We must avoid these behaviors at all costs, because they will suck the passion and drive out of our people.

LEAVE A LEGACY

As we conclude our time together, let's remember it is our responsibility to leave a positive impact on everyone around us and everything we touch. Just like an emergency scene, it is our goal to make the fire service, our fire departments, and our crews the best they can be. We may not all be in formal leadership positions, but we can all lead and coach others to success. Helping others achieve success is the true definition and pathway to us achieving personal and professional excellence.

> "Don't leave a legacy in money, fame, or power.
> Leave a legacy in other people's lives.
> If you do that, it will be worth everything."
> - Tim Tebow -

ACTION STEPS

1. Who has been a coach or mentor to you? What qualities made them a successful coach? Adopt and emulate these qualities.

2. Find someone to coach you. It does not have to be a formal coaching/mentorship relationship. Keep it casual and real.

3. If you are an officer or a senior firefighter, identify another firefighter who you can coach.

4. Finish this sentence: *"When I leave the fire service, I want my legacy to be: _____."*

EVERYONE LEAVES A LEGACY. WE WILL LEAVE LEGACIES WORTH FOLLOWING.

FirefighterSuccessBook.com

ADDITIONAL RESOURCES

FREE DOWNLOAD:
FirefighterSuccessBook.com/SpecialReport

ADDITIONAL RESOURCES

Listen to and learn from successful firefighters and fire service leaders on the *Firefighter Success Podcast*.

FirefighterSuccessPodcast.com

Firefighter Functional Fitness is the essential guide to optimal firefighter performance and longevity. The *#1 Amazon Bestseller* has helped improve the health of firefighters in almost 50 countries worldwide. Authors Jim Moss and Dan Kerrigan provide firefighters with a simple, yet comprehensive approach to improving their fitness, reducing their risk of line-of-duty death, and enjoying healthier retirements.

FirefighterFunctionalFitness.com

Instagram | Facebook | Twitter - @FirefighterFFit

ADDITIONAL RESOURCES

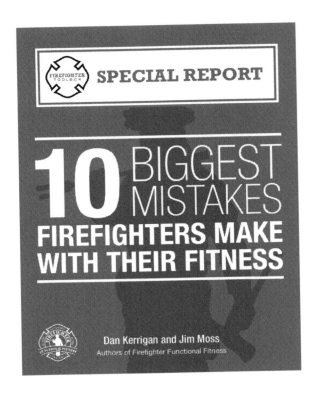

FREE DOWNLOAD:
FirefighterFunctionalFitness.com/SpecialReport

ADDITIONAL RESOURCES

Firefighter Preplan is the ultimate guidebook for thriving as a firefighter. Author David J. Soler shares the strategies and tactics of great and respected firefighters and a plan for how every firefighter can become one.

FirefighterPreplan.com

ADDITIONAL RESOURCES

Firefighter Toolbox is a resource for firefighters and leaders who want to learn and train to maximize their God-given potential. Firefighter Toolbox provides articles, training bulletins, and podcast episodes for firefighters to take their skills and knowledge to the highest level.

FirefighterToolbox.com

ABOUT THE AUTHOR

Jim Moss is a fire captain and paramedic for the Metro West Fire Protection District in St. Louis County, Missouri. His fire service passions include leadership, training, and mentorship. Along with Chief Dan Kerrigan, Jim is the co-author of the #1 Amazon Bestseller: *Firefighter Functional Fitness*, available at FirefighterFunctionalFitness.com.

Jim is a contributor to *Fire Engineering Magazine* and FirefighterToolbox.com. He has shared his message at FDIC International, Firehouse, the International Society of Fire Service Instructors, the National Volunteer Fire Council, International Association of Fire Chiefs, and with fire departments nationwide.

Contact him via email at jim@firefightersuccessbook.com. Connect with him personally on Facebook, Twitter, LinkedIn, and follow him and *Firefighter Success* on the following social media outlets:

Instagram: @FirefighterSuccess and @FirefighterFFit
Facebook: @FirefighterSuccess and @FirefighterFFit
Twitter: @FireSuccessBook and @FirefighterFFit

FirefighterSuccessBook.com

ABOUT THE PHOTOGRAPHER

Chris Smead started CSmeadPhotography in 2008 as a way to document the dedication and hard work of the San José Fire Department through photography. In 2013, his photography started to branch out from local departments as he traveled around California to capture large scale events such as the Rim Fire, the Loma Fire, the Tubbs Fire, the San Jose flooding, the California Camp Fire and, most recently, the Gilroy Garlic Festival shooting.

Chris' work has been featured internationally on multiple occasions and has been used for the training and archives of several fire departments. His work has been seen in publications such as the *Washington Post, Reuters, Japan Times, NBC,* and *France 24.*

Chris is trying to change the common misconception that firefighters do more than put out fires and play basketball with Dalmations. Whether it is a school visit to teach kids about fire safety, a mutual aid fire in the hills, or a rescue call in the middle of the night, he is always at the ready to showcase the true work of firefighters and connect the public with these heroes.

Chris Smead
CSmeadPhotography.com
@CSmeadPhotography - Instagram, Facebook and Twitter